Five centuries of map printing

Published for the Hermon Dunlap Smith Center
for the History of Cartography
The Newberry Library

Series Editor/David Woodward
Previously published
Maps: A Historical Survey of Their Study and Collecting
by R. A. Skelton (1972)
British Maps of Colonial America
by William P. Cumming (1974)

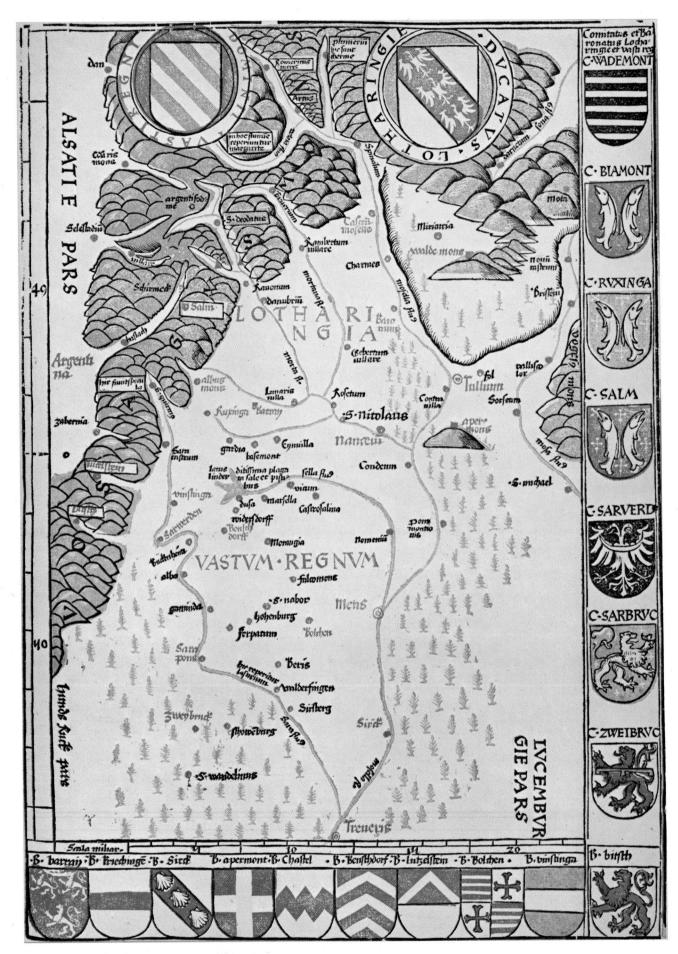

[Lotharingia], in Claudius Ptolemaeus, *Geographia opus,* Strassburg, Johannes Schott, 1513
Courtesy of the Ayer Collection, the Newberry Library

Five centuries
of map printing

Edited by David Woodward

The Kenneth Nebenzahl, Jr., Lectures in the History of Cartography at the Newberry Library

The University of Chicago Press · Chicago and London

ELIZABETH M. HARRIS is assistant curator,
Division of Graphic Arts, of the Smithsonian Institution.
C. KOEMAN is professor of cartography,
Geographical Institute, Utrecht.
WALTER W. RISTOW is chief of the Geography
and Map Division of the Library of Congress.
ARTHUR H. ROBINSON is Lawrence Martin Professor of Cartography,
the University of Wisconsin.
COOLIE VERNER is professor of adult education
at the University of British Columbia
DAVID WOODWARD is curator of maps and program director
of the Hermon Dunlap Smith Center
for the History of Cartography at the Newberry Library.

The University of Chicago Press, Chicago 60637
The University of Chicago Press, Ltd., London

Library of Congress Cataloging in Publication Data

Main entry under title:
Five centuries of map printing.

   (The Kenneth Nebenzahl, Jr., lectures in the history
of cartography at the Newberry Library)
   "Published for the Hermon Dunlap Smith Center for the
History of Cartography, the Newberry Library."
   Bibliography: p.
   Includes index.
   1.   Map printing—History—Addresses, essays,
lectures.   I.   Woodward, David, 1942–   ed.
II.   Hermon Dunlap Smith Center for the History of
Cartography.   III.   Series.
GA150.F58   686.2'83   74–11635
ISBN 0–226–90724–4

Contents

Editor's note

The essays published here stem from six invited lectures comprising the Third Series of Kenneth Nebenzahl, Jr., Lectures in the History of Cartography at the Newberry Library, held at the library on 2–4 November 1972. They were planned from the beginning as a book, around the theme *Five Centuries of Map Printing*. Professor Robinson's paper serves as an introduction, examining the changing relationship between the map printer and the cartographer throughout the history of map printing. Dr. Woodward, Professor Verner, and Dr. Ristow provide a summary of the three major historical methods of printing: woodcut, copperplate engraving, and lithography. Dr. Harris has written on the experimental processes of the late eighteenth and nineteenth centuries, and Professor Koeman brings us up to date with his treatment of the application of photography to map printing and the transition to offset lithography, the technique now in routine use for all but the most specialized classes of maps.

This book does not claim to be a complete history of map printing. While we have striven for editorial consistency, and while the topics for the original lectures were chosen with an eye to a balanced treatment of the topic, the basic approach to each chapter remains each author's own. There are therefore some obvious overlaps and some gaps in the story. For many aspects of the study of historical map printing methods, the field is untouched. If this book serves as a springboard for further research in this neglected area, it will have achieved its major purpose.

The volume is the product of the inspiration and effort of many people, and it is a pleasure to acknowledge them here. The idea of a book on the history of map printing was the subject of much correspondence between the late R. A. Skelton and myself some years ago. When I came to the Newberry Library in 1969 it seemed a natural topic for the 1972 Kenneth Nebenzahl, Jr., Lectures, commemorating as it would the five-hundredth anniversary of the first map printed in the Western world (Augsburg, 1472). I began to look for contributors to such a volume, as it seemed unlikely that any one person would have the required expertise in such an undeveloped field. My thanks are due to the contributors I found, who cheerfully met the varied and numerous deadlines both for the lectures and this book. I would also like to acknowledge the help of many friends and colleagues for help with typing, supplying illustrations, and all the other small but important details of production. Many individuals and libraries loaned material for the exhibition accompanying the lectures; illustrations of many of these items are included in this volume. While they are individually recognized in the captions, I should like to thank them collectively here. Personal thanks are due to my wife Rosalind for help with the index and the proofs; to Lawrence W. Towner, James M. Wells, and Robert W. Karrow, Jr., of the Newberry Library for constant encouragement and support during the lectures in 1972, and to Mr. and Mrs. Kenneth Nebenzahl, whose generosity to the Kenneth Nebenzahl, Jr., Fund has made this and previous books in the series possible.

David Woodward

Illustrations

1 | Mapmaking and map printing: The evolution of a working relationship

Arthur H. Robinson

The first printing, five hundred years ago, in 1472, of a simple, allegorical map was an important event in the history of communication (fig. 1.1). It ranks a close second to that of printing from movable type, a development that occurred only some twenty years earlier. To appreciate properly this momentous event, we must remember that in all preceding time maps had existed only in manuscript form. That basic fact allows two important assertions: first, there could be only a few maps and, second, one could never be sure whether the content of a map was the work of the original maker or merely reflected the independence, or carelessness, of a copyist. Obviously, both inhibited scholarship.

The capability of printing maps immediately opened the way for countless numbers of exact duplicates that, for the first time, allowed scholars easily to compare many geographical portrayals, consider their characteristics, and plan ways to produce even better images of the emerging world. No doubt it also had a very considerable psychological impact on mapmakers, since the realization that their work could be widely subject to critical review probably served as an incentive to some and an inhibition to others.[1] To the age-old art and science of mapmaking a tremendously significant new element had been added—the printer.

Rather, as in the production of a child, the printed map requires the cooperation of two contributors who differ in essential ways. As R. A. Skelton has pointed out, both the printer and the cartographer became professionals rather suddenly in the late fifteenth century,[2] and because it has always been quite uncommon for one person to combine both activities, these two graphic artists have had to function together most of the time. Although they have worked side by side for five hundred years, their association has changed considerably along the way. The objective of this essay is to survey this alliance, to try to characterize the evolution of what has been an essential

Note: It is obvious that any historian must depend heavily on the work of others, and I am no exception. I would particularly like to acknowledge the value of many discussions with the late R. A. Skelton of the British Museum. I owe an especial debt to Dr. Helen Wallis, also of the British Museum, and the contribution of two former students, David Woodward and Karen Pearson. Our many conversations and exchanges and their bibliographic finds relative to the history of cartographic technology, the cartographic applications of cerotyping, early lithography, and the use of type on maps have been invaluable.

partnership between the cartographer and the printer, to examine the technical foundations upon which this relationship has been based, and to look at some of the ways the changes through time have affected the character of the printed map.

Students of intellectual history emphasize quite properly the importance of the invention of printing and the flood of information it let loose on the world, but usually they refer only to the printed word obtained from movable type. One thing, not generally appreciated, is that a graphic image, which of course includes the map, is quite a different thing from the printed word in some significant respects. For much longer than five hundred years the marks that comprise the letters of the alphabet have been thoroughly familiar to the literate. Consequently, unless the print is outstandingly ugly or difficult to read, it is ordinarily quite transparent to the reader in the sense that he absorbs its message without much reaction to its graphic characteristics. On the other hand, a map employs lines, shading and tones, patterns, some words, and all sorts of other kinds of complicated and generally unfamiliar marks. A map is an infinitely more complex thing from a perceptual point of view, and its combination of marks is not at all transparent. Consequently, while the viewer is absorbing the information contained in the display, he is also reacting in important ways, both consciously and unconsciously, to the graphic characteristics of the map.

Because of the conceptual complexity of a map, the mapmaker must not only position his symbols with care in order to make his map accurate, but he must also pay close attention to its total graphic design. Both the symbology and the entire display need to be planned carefully in order to evoke the right total image as well as to arrange the graphic priorities in proper order. Furthermore, the cartographer has the added responsibility to make the whole thing attractive enough not to repel the potential user, for maps tend to confuse many people. It is because of these fundamental factors that the working relationship between the printer and the mapmaker assumes a critical importance, for during most of the last five hundred years the printer often has had as much or more to say about these graphic matters as has the cartographer.

Before beginning the examination of the evolution of the critical association between the cartographer and the printer, it is necessary to explain why an essay on this subject is appropriate. One might reasonably expect that the matter had already been thoroughly investigated, but it has not. There are at least two important reasons why.

The first is that it is difficult to acquire the necessary information, although that is not a very scholarly excuse. There have not been many careful studies of cartographic practice in past centuries, that is, studies of how the cartographer actually went about his science and art, how he compiled the data, the mechanics of tracing and construction, the preparation of the drawing for duplication, the proving procedure, and so on. Bits and pieces abound, and there are occasional details, but thorough assemblies and analyses are both lacking. Furthermore, as Ivins has pointed out in his excellent book, *Prints and Visual Communiation,* written descriptions of everyday techniques did not appear until long after the methods had been developed.[3] For example, although letterpress printing, engraving, and woodcutting have all been practiced from the fifteenth century on, the first written technical account of the tools and processes employed in letterpress printing did not appear until more than two hundred years later, in 1683. The first description of the tools and techniques of engraving is dated 1645, and the first account of woodcutting came only in 1766, more than three centuries late, so to speak. If we are so relatively uninformed about the activities of both the mapmaker and the map printer, it might seem unwise to attempt to survey the evolution of their working alliance, but there is value in the general overview. The telescoping of time sometimes allows us to obtain a perspective on important trends and changes that are often not so apparent in detailed studies.

The second reason why this topic is not well known is that most students of the history of cartography seem to have had little interest in the actual methods involved in mapmaking. They have tended to concentrate on the map as a record reflecting geographical understanding and have paid little attention to its graphic characteristics. We have finally come to realize that the graphic design of maps is as vital a part of cartography as is literary style in the world of letters. Certainly, there is a strong association between what we can call the look of maps and the actual technology of their preparation.

One final introductory point is that we need to clarify what we mean by the terms "cartographer" and "printer" as they will be used in this essay; neither is easy to define precisely. By the term "cartographer"—or "mapmaker"—we shall refer collectively to the person, the institution, or simply the operations involved in planning and compiling the map, in deciding the area to be covered, what names to be incorporated, the geographical data to be shown, the kinds of symbols and legend to be employed, and so on. By the term "printer" we shall refer collectively to the individual, the agency, or simply the activities primarily concerned with duplicating the map, in the preparation of the printing surface, the operation of the press, the making of corrections, and so on. Sometimes, as we shall see, these activities become confused in practice, but the general distinction is sound.

We shall begin our survey of the evolution of the working relationship between the mapmaker and the map printer with the hand-drawn map not intended for printing. The author of a manuscript map naturally labors under some graphic constraints. It is true today and it was true a thousand years ago. These constraints include his own understanding of the intricacies of graphic perception, his manual ability, and, of course, the limitations of the media with which he works whether they be vellum or paper and the reed, steel pen, or brush. Within limits he can attack the constraints by learning more about perception, by practicing, or changing his medium. Other than the sizes of skins, paper, or walls upon which to draw, he really does not have much in the way of restraints imposed upon him. Cartographically speaking, for better or worse, the manuscript mapmaker is his own master.

When the printing of maps became possible, a new element was introduced that caused a great change in this very intimate connection between the cartographer and his map. The opportunity of obtaining numerous copies exactly alike obviously made it desirable for many maps to be duplicated. However, to take advantage of this development, the cartographer had to submit to the controls imposed by these printing processes. These constraints constituted a kind of toll or tax levied for the privilege of duplication, and they were exacted in units of versatility. These restrictions varied, depending upon the process, but they were in all cases significant. Some were constraints of a general nature such as the practicality of employing color or the much more serious problem of how to accomplish the necessary lettering on the maps. Some were more specific such as those having to do with the size and character of lines and individual symbols or the problem of how to produce the tones and patterns so necessary to the development of area symbolism, the desirable figure-ground differentiation, and the creation of the impression of a third dimension. Five hundred years ago these constraints suddenly became matters of vital importance to the cartographer, and they have been of constant concern ever since.

There is, of course, another side to the coin. As David Woodward has observed, maps have a variety of characteristics that have placed unusual demands on the printer.[4] Their desirable size is often a problem. Commonly their geometric fidelity demands a dimensional stability hard to obtain. Maps frequently need revision. They may be produced as separates, but it may be advantageous to incorporate them with text material. In general, maps require a fineness of line, a level of consistency, and a number of copies that are often at odds with one another. Certainly the problems placed upon the mapmaker by the printer were matched by the problems placed upon the printer by the cartographer.

We shall begin our survey of the evolution of the working relationship between the mapmaker and the map printer by attempting to divide the five hundred years since 1472 into periods that seem to differ from one another in some significant respect. We can easily recognize at least four as presented in diagram 1.1. The first century, ending around 1570, was a confusing one of trial and error involving the development of two competing printing methods and during which cartographers and printers were getting acquainted and feeling out one another. Around the end of

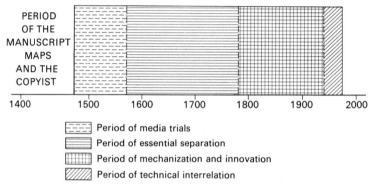

1.1 Periods in the relationship between the cartographer and the map printer

the sixteenth century a second period began to develop during which the cartographer and the printer settled into a formal and straightforward but rather standoffish kind of association that lasted for more than two hundred years. Toward the end of the eighteenth century the mechanization associated with the Industrial Revolution began to creep into map printing and together with some other innovations inaugurated a third era. As pointed out by Kenneth Nebenzahl, the nineteenth century differed markedly from the earlier periods, being characterized by great change and variety in printing methods.[5] Things are different today, in the fourth period. Although we are a little close to it to be able to obtain an excellent perspective, it is likely that a new period came into being around 1940. During the latter part of the third period the relationship between the printer and the cartographer gradually ceased to be disjunctive and now in the fourth period it is becoming essentially symbiotic.

The variations in the evolution of the working association between the cartographer and printer are directly related to the basic methods of duplication and to the development of associated techniques such as electroplating, transfer, photography, and so on. In this evolution the technology of printing seems generally to be a good deal more important than the mapping and cartographic techniques. Accordingly, we shall first describe briefly the essentials of the major methods of duplication used for maps, and their cartographic consequences. Then we will outline the history of the relative employment of these methods. Finally, as examples of the critical nature of the relationship between the cartographer and the printer, we shall examine the changes through time of the degree to which the printing technology allowed the cartographer to control the ultimate result.

We will restrict our consideration of printing to those repetitive methods wherein a fluid ink is first applied in a desired pattern to a printing surface and then transferred to paper. That restriction eliminates many methods of duplication that, of course, can be used for maps such as photocopy, diazo, electrostatic—which includes the ubiquitous Xerox copy—spirit-gelatine, silk screen, and perhaps some other methods used for comparatively small numbers. Sometimes very useful in specific instances, they are not at all important in the history of cartography, although conceivably they might one day become so. With this restriction one can recognize what we might call the generic types of printing: there are three—relief, intaglio, and planar, the last also called planographic or simply surface printing.[6] We shall describe them only briefly.

Relief printing, probably the oldest of the three, requires a two-level surface, the outer or upper of which carries the image and receives the ink, and a lower surface that does not, simply by reason of its being at a different level. When pressed on paper the ink on the outer surface is transferred. First used in the form of carved wood blocks for stamping designs, much later on it was accomplished with metal plates affixed to wood. Relief printing is the method employed by Gutenberg and the others who developed movable type in the middle of the fifteenth century, on account of which it is often simply called letterpress. A two-level relief printing surface can be obtained in many ways.

There are some problems inherent in relief printing that are significant to the mapmaker. First, a smooth paper is essential for good quality and intricate images. Second, an individual printing is made by a single impression, and each impression requires a separate inking of the image area. Consequently, only the amount of ink to be transferred to the paper each time should be applied to the raised surface, and the printing pressure must be just right, since any excess ink will creep over the edges of the raised surface or the shoulders and into the non-printing areas that are supposed to remain white in the print. If ink does build up in this way, the printed lines and marks tend to become coarser and the intervening white spaces smaller.

Intaglio printing is entirely different. In this process one starts with a smooth, polished surface, almost always copper—a comparatively soft material—and by engraving or gouging creates a recessed image in the metal. The incisions may range from furrows for lines to tiny individual pecks and nicks. Ink is then spread over the surface and into the recessed image areas, after which the surface is wiped clean, with, however, the ink remaining in the depressions. When dampened paper is placed on the copperplate and great pressure applied, the paper is forced into the depressions and when it is lifted, the ink adheres to the paper. Intaglio is just the opposite of relief printing since the image is carried at the lower level, so to speak. Although the basic process is usually called copperplate engraving, the incised image can be produced in other materials and in other ways, but almost all map engraving has been on copper.

As in relief printing there are some inherent problems in the intaglio process that are significant to the mapmaker. A primary characteristic of copperplate engraving as a duplicating process is that the production of copies is comparatively slow since it takes time to prepare, ink, wipe clean, and then apply the great pressure needed to transfer the image. Furthermore, the copperplate wears easily under the great pressure and can only produce a limited number of prints.[7] One the other hand, ink build-up and spread is not the problem it is in letterpress printing since the plate is wiped clean after each inking; consequently, fine lines and intricate detail can easily be obtained and maintained.

While both the relief and intaglio processes depend upon an actual physical separation between the image and the non-image areas, planar printing accomplishes this by what can loosely be called a chemical separation. In this process the image area on some smooth surface is made receptive to ink while elsewhere the surface is given the ability to prevent the adherence of ink. There are several ways to accomplish this separation, depending upon the substances involved, but by far the most common in times past and today is based upon the fact that oil or grease and water tend to reject one another. If a drawing is made with a greasy substance, and if the surface can be alternately dampened and inked, then the ink will adhere only to the oily places and the water only to the non-oily parts. Paper pressed against the surface, therefore, will pick up an image only from the inked area.

The smooth surface first used was that obtained on a close-grained limestone block obtained from Solenhofen in Austria. Sometime later, the process became known as lithography, literally rock writing, a name it still carries, although stone has for some time been replaced by thin metal plates that can be bent around a cylinder for rotary printing. Lithography labors under no significant technical requirements that place constraints on its capabilities such as the need for smooth paper in letterpress or the great pressure required in copperplate engraving.

Each of these three families of printing methods has had varying capabilities for the duplication of the graphic requirements of maps. Consequently, their changing relative importance during the past five hundred years is a matter of great interest in the history of cartography, and diagram 1.2 is an attempt to portray this. It

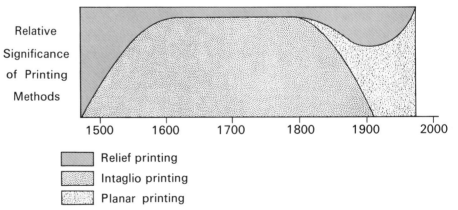

1.2 The variations in the relative significance of the basic map printing methods

is constructed on far more intuitive grounds than is desirable, but as was noted earlier, studies in the history of cartography are generally quite deficient in their treatment of the technical aspects of mapmaking, which, of course, includes the printing.

The apologia should also be extended to the form of the diagram. The vertical width of each zone representing a process is intended to be proportional to the significance of that method during the time indicated on the horizontal axis. One can only guess at the relative frequency with which the processes were employed, but this is clearly difficult, for how does one equate the numerous impressions from a simple three- or four-inch square relief block in a large edition of a printed book with the relatively few impressions from a forty-times-larger copper-engraved topographic map? Furthermore, the vertical sum of the shaded areas is meant to represent the total production of printed maps with no regard to the fact that the total has varied from a few thousand or so in the years before 1500 to many millions in just the past few years. It is easy to quibble with the delineation, but it is meant only to be a broad generalization. The smooth simplicity is intentional and is a consequence of our lack of quantitative knowledge regarding production figures.

The earliest surviving printed map is apparently a small, marginal illustration in the *Etymologiae* of Saint Isidore, bishop of Seville (fig. 1.1). That work, written in manuscript in the first half of the seventh century, was finally set in type and printed more than eight hundred years later in Augsburg, in 1472. The print was produced from a relief woodcut, and although the map is little more than an allegorical diagram, it exhibits some of the characteristics that doomed the early relief woodcut in competing with intaglio copper engraving, namely a coarseness of line, an enforced angularity, and great difficulty with the lettering, especially the smaller. Another major difficulty of the woodcut technique was the production of tone. Parallel lines were about all that was possible, although cross hatching was occasionally done. Stippling, that is, many small dots, was extremely difficult.

The early woodcut map was fashioned from a plank, that is, a slice of wood sawn with the grain. The white areas of the map were made by cutting them out, leaving only the image area standing in relief. The woods commonly used were apple, cherry, beech, box, or such other varieties that were not greatly affected by the gums, oils, varnishes, and water used in inking and cleaning the blocks.[8] The sizes of maps that could be handled, of course, were limited by the sizes of the presses, the paper, and the wood planks. Large maps had to be assembled from prints from separate blocks such as the one naming America for the first time, the 1507 map by Martin Waldseemüller of Saint Dié. That heroic map covers nearly 26 square feet, and was assembled from prints from 12 separate wood blocks, each of which was approximately 38 × 53 centimeters or 15 × 21 inches.

Probably the thorniest problem facing the cartographer and the printer employing the woodcut was the problem of lettering, since in woodcuts the whites are carved away to leave the image area standing, and for fine detail and small lettering this is very difficult. This problem spawned several ingenious methods to overcome the inability of the woodcutter to carve small enough lettering of acceptable quality.[9] A rather common technique, which serves to emphasize the difficulty of the problem, is one wherein the printer cut holes right through and wedged the metal type in the slots so that it would print along with the other map detail. Another method was to obtain thin stereotypes, that is, castings from molds of set type, and then glue them to the woodblock.[10] One way or another, small lettering could be incorporated, and had the woodcut technique permitted the other attributes desired by cartographers, it probably would have had a more important place in the history of mapmaking.

Although the woodcut did not inhibit clean, straightforward composition in the broad sense, it simply did not allow either good tonal representation or the requisite fineness of line and detail. Furthermore, the dimensions of woodblocks that were available and could weather the presswork and would not warp or split placed a limitation on the size of maps. Consequently, as the sixteenth century wore on, the relief woodcut block lost some of its favor and the intaglio, copper-engraved plate became the preferred method for printing maps. For a long time, however, woodblocks were widely used for the maps and diagrams in books such as Sebastian Münster's widely distributed *Cosmography,* the 1550 edition of which contained more than nine hundred woodcuts. Woodblock maps and other illustrations could be locked up in the chase along with the text letterpress, and the material for the whole page could all be inked and printed at once. On the other hand, for a printer to combine on one page an impression from an intaglio copper engraving with text from relief letterpress meant that that sheet would have to be printed twice, a slow, costly requirement with the added disadvantage that the impressions had to be obtained from different presses. Nevertheless, as diagram 1.2 attempts to show, by the end of the sixteenth century intaglio printing became the more important method of duplicating the larger, separate maps, and it kept that position for more than two centuries.

Copper engraving offered several advantages over the woodcut, chief of which were that larger plates could be used, that a fine line could be obtained, that small, neat lettering was possible, that smooth curves could be produced, and that clean, fine-textured, flat, and variable area tones could be had, a very important component of cartographic design. The lines and lettering on the copperplate were obtained by pushing the burin or graver forward into the copper to cut a furrow to receive the ink and then trimming off the burr that rose on each side. That this required great skill is evident. To obtain curves such as contours, the plate could be swung around as needed. The incised lines could be as narrow as desired, although wider lines had to be produced by many smaller, closely spaced, parallel grooves.

One of the major concerns of the printer and cartographer is the process of revision, either to correct mistakes in the initial construction or to make changes and bring up to date an older map. To change the printing surface of a woodcut was inherently difficult since it could only be accomplished by carefully removing the section of the block to be revised and inserting a fresh piece of wood. Compared to this, changing a copperplate was easy. Careful hammering on the back of the plate would raise the surface, which could then be ground down, polished, and re-engraved. Although relatively easy correction provided a great saving to the cartographer and printer, it did not necessarily work to the advantage of the map-using public. As Skelton points out: "The life of a copper plate was prolonged, by husbandry and reworking, long after the map engraved on it had outlived its reliability. . . ."[11] As a matter of fact, a few plates had remarkable longevity. Some of Christopher Saxton's county maps of England, first engraved in the 1570s, were still being issued in altered form nearly two hundred years later.[12]

The very real advantages of the intaglio copper engraving over the relief woodcut as a technique for preparing maps for printing, of course, was accompanied by some less desirable attributes. Copper engraving was very expensive. In the first place, the copperplates themselves were costly since they had to be relatively thick and, second, the engraving process was very slow. For an individual map the outlay for copper might not seem excessive, but for a major work it could quickly

giones·quarū breuiter nomina et situs
a paradiso ¶Paradisus est locus in or

1.1 [Tripartite world map], in Isidore of Seville,
Etymologiarum, Augsburg Günther Zainer, 1472
Courtesy of the Ayer Collection, the Newberry Library

1.2 G. H. Dufour, *Topographische Karte der Schweiz*, Bern, 1833–63 [detail]
Courtesy of the John Crerar Library

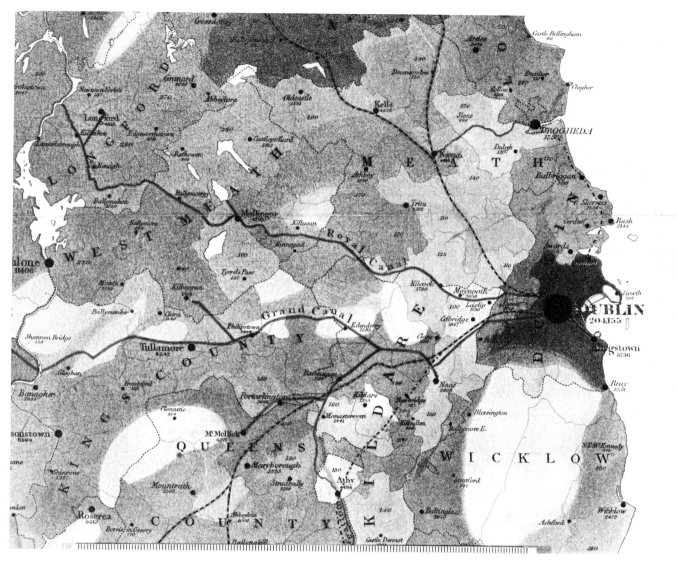

1.3 Aquatint engraving. Henry Drury Harness, "Map of Ireland . . ."
in *Atlas to Accompany the Second Report of the Commissioners Appointed to*
Consider and Recommend a General System of Railways for Ireland,
Dublin, H.M.S.O., 1838.
Courtesy of Professor Arthur H. Robinson

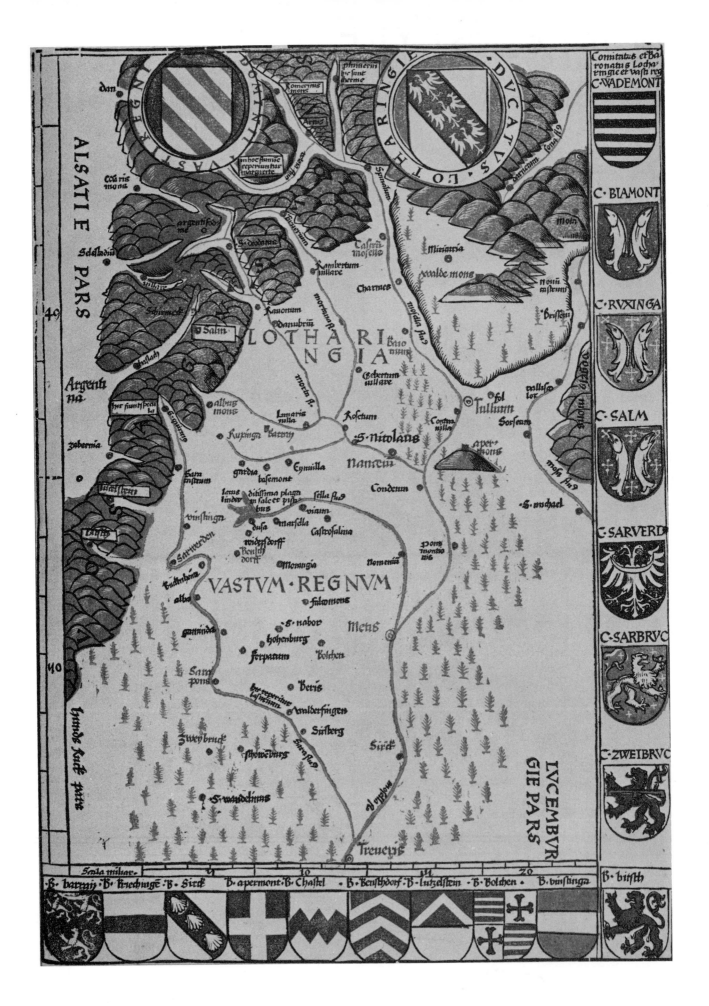

1.4 [Lotharingia], in Claudius Ptolemaeus, *Geographia opus,* Strassburg, Johannes Schott, 1513
Courtesy of the Ayer Collection, the Newberry Library

1.5 Sheet 13, Queen's County, *Ordnance Townland Survey of Ireland,* 6″ = 1 mile. 1841
[detail]

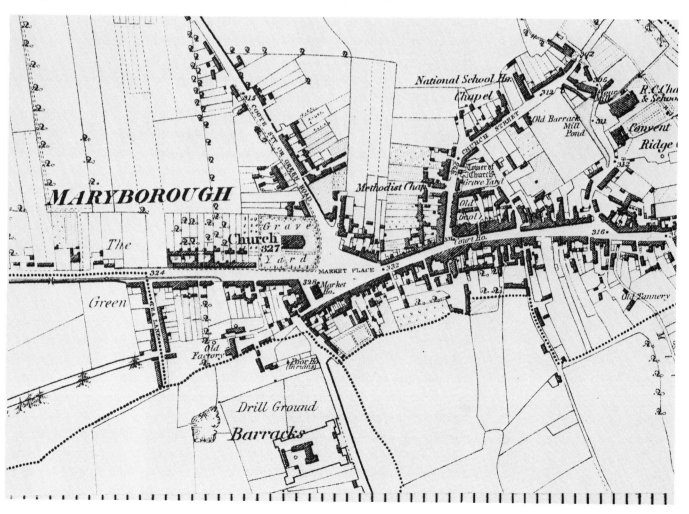

amount to a disquieting total. For example, Lucini, the engraver of the second edition of Robert Dudley's *Dell' Arcano del Mare,* a seventeenth-century sea atlas, is quoted as observing that he used no less than five thousand pounds of copper in engraving the maps and illustrations for it.[13] When the burgeoning, large-scale survey organizations of the latter part of the eighteenth and the early part of the nineteenth centuries were called upon to produce multiple copies of scores of maps, the cost easily could be prohibitive. For example, the Bavarian Survey in the early nineteenth century was faced with the problem of producing several hundred sheets annually and came to the conclusion that it simply could not afford the expense of employing copperplates let alone the cost of their engraving.[14]

Engraving a copperplate took a very long time since it required great skill if done well. For example, Lucini complained that the engraving for the *Dell' Arcano del Mare,* took ". . . twelve years sequestered from all the world in a little Tuscan village."[15] Another example is provided by the celebrated, very detailed Dufour map of Switzerland made in the mid-nineteenth century. An experienced map engraver has estimated that the progress of an individual working on a plate from which that map series was printed would be less than one square inch a day (fig. 1.2). If that is so, then each sheet would have required from one and one half to two years to engrave, and there are twenty-three sheets in the series. The assertion that the cost of engraving a medium-scale map in England would often exceed the cost of the survey to produce it seems reasonable.[16]

In bringing to a close this short summary of copperplate engraving as a medium for the printer and the cartographer, it is appropriate to point out that both the woodcut and copper-engraving processes are essentially line processes. This means that the tonal variations or shading required to develop the figure and ground relationship, to set the land off from the sea, or to render the configuration of the land, or for other symbolism, could employ only line.

Toward the end of the eighteenth century various kinds of mechanical and chemical methods were developed to obtain smoother shading or tonal effects from a copper engraving.[17] Remarkable effects on the copperplate could be accomplished by the use of various sorts of rockers, manipulated either by machine or by hand, or by mechanical ruling. As the need grew for even subtler variations in tone, such processes as aquatint etching were employed. This led to some notable, but not widely utilized, cartographic treatments in the nineteenth century such as the acquatint shading employed on Harness's 1837 map of population density in Ireland (fig. 1.3).[18] Another interesting example of aquatint rendering is the world map of precipitation in Berghaus's *Physikalischer Atlas* of the 1840s, a landmark in thematic cartography.

Interesting as these sorts of ephemeral processes are, of much greater importance in the evolution of mapmaking and map printing was the introduction early in the nineteenth century of the planar or surface printing method of lithography. The potential of lithography as an alternative to copper engraving was realized very early, and it did not take long for the new technique to be put to wide use. The Bavarian Survey, referred to earlier, decided to employ it in 1809 not much more than a decade after its discovery and, as a matter of fact, employed the inventor, Senefelder, as one of its lithographic inspectors.[19] Lithography was a relatively inexpensive process with a great deal of versatility, and it soon began to rival engraving. Copper engravers turned to the new medium and were capable of turning out first-class work since the techniques required were similar to those employed in copper engraving, but were much easier. Senefelder claimed that the same things could be done three times faster on stone than in copper.[20]

It is important to point out here that for quite some time the lithographers tried to make their maps look like engravings, and it was difficult to overcome the conservatism that continued to favor the acknowledged high quality of the copper engraving. For example, the Société de Géographie de Paris, soon after its founding in the early 1820s, set up a committee to study the matter.[21] They concluded that copper engraving was the best method and would never be superseded, but that lithography was suitable for large-scale maps that did not require the fine line possible with copper engraving. Nevertheless, the versatility and quality of lithography steadily improved, and with the perfection of mechanical aids such as grained paper, machine ruling, and the transfer processes lithography began to forge ahead at the expense of copper engraving. Diagram 1.2 shows that today lithography is by far the dominant method, and this is true however one might measure it since books are also being increasingly produced by offset lithography. The figure also attempts to show that during the nineteenth and early twentieth centuries there was a considerable amount of competition among the three printing processes. If one erects an ordinate at about 1890 the proportions would be about equal.

It is in this important respect that the period from about 1800 well into the present century was quite different from the preceding periods. This is true in terms of both the number and the changing importance of the several basic printing methods, although the simplicity of the figure masks the fact that several new ways of producing those surfaces were devised. It is appropriate that we develop a few basic generalizations about that very important period in the history of cartography.

First, we may observe that at no other time in the history of cartography have so many innovations occurred of significance to the printing of maps, innovations that directly or indirectly had a very considerable influence on the graphic design of the map. Within the space of less than a century such developments as photography, wax engraving or cerotyping, electrotyping, line engraving, wood engraving, and the halftone screen were invented and put to the test, to mention only a few. The analogy may be a bit strained, but one can liken map printing in the century and a half after 1800 to a long musical medley. Throughout the period, first one and then another new idea for map printing, like a fresh melody, would come forward and flourish for a while, only to fade out or blend into yet another. Always, however, there was a steady move toward greater versatility.

A second generalization is that, whereas during the previous period the relationship between the printer and the cartographer seems relatively straightforward, the nineteenth and early twentieth centuries were a time of very considerable confusion in this relationship, partly because of the plethora of available methods for duplication, but especially because the various processes were no longer unique and independent of one another as had been the relief woodcut and intaglio copper engraving. Very shortly after lithography was invented the transfer process was developed.[22] In this process a right-reading image could be drawn with special ink on special paper and then transferred face down to a stone. It was as profound a development at that time as is the Xerox copying process of today. Within two decades of the invention of lithography, Senefelder himself observed: "At Munich . . . Paris . . . St. Petersburg . . . all resolutions . . . etc., agreed to in cabinet meetings are written down on paper by the secretary with chemical ink; in the space of an hour, fifty impressions may be had and distributed at pleasure."[23]

Cartographically, the significance of that development was that not only could a map be drawn right-reading on paper and be directly reproducible, but an impression in the appropriate "chemical ink" could be pulled from a copper- or wood-engraved map and then duplicated by lithography. For example, the one-inch map of Ireland by the Ordnance Survey was first engraved on copper during the nineteenth century and printed in black. Late in the century the engravings were revised, separated into colors, transferred to lithographic stones, and printed in color by lithography. The student of the history of cartography often finds it very difficult if not impossible to tell whether a map was actually engraved on copper and then transferred to lithographic printing or whether it was a lithographic engraving to begin with. Furthermore, lithographers became expert. The process of covering the stone with a protective ground and then using a needle to open the ground to receive ink and thus to provide a printing image could produce detail and a fineness of line to rival copper engraving.[24] In these and many other ways the nineteenth and early twentieth centuries turn out to be an incredibly complex period with respect to cartographer-printer relationships.

A third important generalization about the period beginning around 1800 is that the development of the paper-making machine and the steam press in the early nineteenth century opened a very much wider audience for the cartographer. Mass production of printed text material inevitably entrained maps. Until relatively recently, however, that mass production was produced by the relief letterpress, and this required that the incorporated maps also be on relief printing plates. On the other hand, separate maps such as those being produced by the prolific national surveys were more economically printed from the surface lithographic plate. As we have seen, the cartographic capabilities of the two printing processes are quite different. Although David Woodward has documented with care the effect on map design of one of the relief processes, cerotyping, other such studies in depth are lacking.[25]

A fourth generalization is that in the nineteenth century the proliferation of printing processes, led by lithography, made possible the common use of color in the duplication of maps for the first time in the history of mapmaking. To be sure, we know of at least one map that was printed in more than one color from separate wood blocks as early as the beginning of the sixteenth century (the 1513 Strassburg edition of Ptolemy; *frontispiece* and fig. 1.4) and even a three-color process employing intaglio mezzotint plates for red, yellow, and blue inks was invented around the beginning of the eighteenth century, although apparently it did not progress beyond the experimental stage.[26] Nevertheless, the expense and the problems of ensuring that the colors printed in the right place, called registry, kept map coloring a manual process until the dramatic shift to common color printing that came about a little more than a century ago.

Other generalizations could be made having to do with the upheaval occasioned by the numerous innovations in printing methods that were introduced in the nineteenth and early twentieth centuries, ranging from the development of gravure and the introduction of offset to the incorporation of photography in the cartographic and printing process, but space is limited and other matters demand attention. Suffice it to reiterate that the nineteenth century initiated a period of intense experimentation and innovation in the printing methods which had profound effects on cartography.

The twentieth century and the past few decades finds a continuation of this rapid change, but there seems to be a subtle variation in the emphasis. Whereas the major developments in the nineteenth century seem to have been primarily in the area of printing methods, the pendulum of change in the twentieth century has swung toward the techniques of the cartographic process. The modern printer and the modern cartographer are today probably working more closely together as a team than ever before. Naturally there are those who prepare maps for the printer in old-fashioned ways, and there are printers who are ultraconservative, but in modern mapmaking it is often difficult to tell when cartography ends and printing begins.

One other modern trend is worth commenting on. For the majority of the past five hundred years the printer has found it difficult to answer the cartographer's demands for size, precision, expressive symbolism, quality, number of copies, and so on. In the past century the shoe seems to have slipped on to the other foot, and the cartographer is now struggling to find ways to take advantage of the versatility of the printer.

This rather quick glance at the evolution of the techniques employed by the cartographer and the printer to duplicate maps, and their general graphic consequences, has dealt so far with rather basic technical matters. It is important to observe, however, that this complex evolution incorporates a host of more specific topics closely related to printing that merit extended review such as the changing methods of lettering the map, the vital matter of the rendering of tones, or the total effect of the printing method on the graphic design. Although these topics are of basic importance, because of their complexity they are more appropriate as the subjects of separate studies. Accordingly, the remainder of this essay is devoted to what may well be an even more fundamentally important matter in the history of cartography, namely the degree to which the changing working relationship between the mapmaker and the map printer permitted the cartographer to control the graphic characteristics of the printed map. To put it more simply, what printer-cartographer relationships were involved in setting the design of the map, or more bluntly, who decided what the look of the maps would be? Although really quite different, there is in this a kind of parallel with the concern of the student of the graphic arts for the autographic characteristics of the various media, that is, the degree to which the tools and techniques enable the artist to express himself freely. This is clearly an important topic, but it is a difficult subject to document.

Let us begin by characterizing the essential tie between the mapmaker and the printer as consisting of what is often called the map worksheet or manuscript drawing made by the cartographer. The worksheet is the basic compilation drawing that establishes the major characteristics of the map. The printed version of the map is prepared, more or less directly, from the map worksheet, with its accompanying written specifications for line weights, tones, lettering, and so on. In woodcut and copper engraving, this meant that a separate craftsman had to follow the worksheet. Today various kinds of artwork are prepared by specialists or draftsmen using the worksheet as a guide. It follows that a comparison of the printed map with the worksheet should enable one to judge how well the construction and duplicating processes have fulfilled the intent of the cartographer. Unfortunately, worksheets all too often have been used up, so to speak—one might be pasted face-down on the woodblock to guide the cutter—and in any case they are not often kept, even today. Apparently very few have survived. Consequently, much of the earlier part of this evolution must be based largely on speculation and inference.

It is probable that few if any cartographers were also woodcutters in the early period (diagram 1.1). In the strict medieval guilds, woodcutters were classed with carpenters, and were highly specialized craftsmen.[27] It seems safe to assume, then, that the vast majority of woodcut maps were prepared, not by cartographers, but by the printers who often relied upon a number of these more or less independent technicians, or even the itinerant journeymen of which we have some evidence.[28] What changes took place in the transformation of the cartographer's worksheet to the ultimate woodcut block we can only imagine, but they were likely to have been considerable. There is extant, however, one manuscript drawing that might provide a clue.

In 1516 Martin Waldseemüller produced a large map called the *Carta Marina.* Although this map and the 1507 map referred to earlier had long been known, all copies seemed to have been lost until one was discovered by J. Fischer late in the nineteenth century in the library of the castle of Wolfegg in Würtemberg. These are wallmap-like productions, each sheet printed from one of twelve woodcut blocks. The prints of the 1516 *Carta Marina* specimen are thought to be proof copies taken from the blocks because the prints overlie a network of red lines representing the graticule of latitude and longitude. The second sheet in the middle row, covering north and west Africa, does not happen to carry this grid. In addition to the prints, the codex also contained a manuscript version of this particular sheet. The editors of the facsimile edition, Professors Fischer and Wieser, assumed the manuscript sheet to be a tracing of a print from the original block.[29] It seems entirely possible, however, that this manuscript version may instead be the compilation worksheet or drawing intended as the original guide for the woodcutter. In any case, a comparison of the two reveals numerous differences, many of which seem regressive, suggesting that it is either a very bad tracing or that it might be the original compilation drawing. For example, if it were for purposes of correcting the block there would be no need to do such a thorough but sloppy job of tracing.

It is not my purpose here to argue one way or another regarding the real purpose of the manuscript version of that sheet, interesting as the question is. Clearly, how-ever, there is a vast difference between the two, and it is reasonable to assume that the discrepancies are representative of the kinds of differences that might have occurred between a cartographer's plans and a woodcutter's rendition. The shading, the drawings, the positioning of the lettering and various other elements of a map were at the mercy of the craftsman. In the early period it is difficult to imagine that a woodcutter would follow directions closely as to line widths, letter sizes, tonal darkness, and the other graphic characteristics of a worksheet. Woodcutters were probably an independent breed.

It is reasonable to assume that the printer-cartographer relationship in the early days did not promote the autographic aspect of the cartographer and his map. It is only fair to add, however, that it is not unlikely that many woodcutters were better graphic artists than were many cartographers, and that the lack of the autographic quality in the relationship may have been a blessing in disguise.

The situation with respect to engraving was no doubt much the same, in spite

of the fact that some cartographers were also engravers, the most notable example being the great European cartographer Mercator.[30] Copper engraving was a far more versatile medium, and the ability of the skilled engraver to produce sharp delicate lines, expressive tonal variations, and lettering ranging from the tiny and precise to the flowing and elegant no doubt led the engraver increasingly to dominate cartographic design. For example, at this stage of our understanding of these matters, one can only speculate how much of the balance and decorative standard of John Speed's maps of England's counties may have been directly contributed by the engravers of Holland, such as Jodocus Hondius, where Speed sent his maps to be engraved.[31]

It is doubtful that the occasional mapmaker, presumably awed by an engraving tradition that had acquired the stature of age, would have been brash enough to have specified line widths and spacings or lettering styles and sizes in other than general terms. Furthermore, in the earlier days of engraving the gulf between the individual cartographer and the engraver-printer may have been widened by the fact that engraving was apparently a home activity. Cornelis Koeman points out that the Blaeu mapmaking-printing establishment in Amsterdam, which contributed both major cartographic works as well as an innovative press design, probably had no engraving department.[32]

On the other hand, the professional mapmaker no doubt was a regular employer of engravers and in his manuscript compilations he probably anticipated the general designs. Clearly, engravers could and did produce faithful copies of the manuscripts they were given from which to work, but it is also clear that the manuscript worksheets were prepared in the first place in what could be called "an engraving style" set by the traditions that had developed because of the characteristics of the media.

In the latter part of the eighteenth century and in the nineteenth century the growth of national surveys no doubt led to the development, in institutional mapmaking, of a much closer kind of relationship between the cartographer and the printer since the survey establishments did employ "in-house" staffs of engravers. For example, in 1820 the Ordnance Survey office at the Tower of London employed one principal and three apprentice engravers, and in the early 1830s there were eleven on the staff employed at the Ordnance Survey office in Phoenix Park in Dublin to engrave the six-inch maps of Ireland.[33]

The cartographic design of such maps had no doubt been largely chosen from among the alternatives that had been devised by engravers who had practiced the craft for centuries and passed it down through the apprentice system. One has little trouble recognizing an engraving, which if done in the so-called best style is characterized by very fine lines, delicate stippling and lining, and a general appearance of elegance. Sometimes the fineness of line and the tiny lettering seem incredible. Sir Charles Close in his *History of the Ordnance Survey* provided a most revealing comment on contemporary cartographic taste and graphic priorities in referring to the six-inch series of Ireland when he wrote, "The maps were beautifully engraved; indeed, the engraving is so fine that it is difficult to reproduce it" (fig 1.5).[34]

Although institutional maps probably came out the way they were planned, the private mapmaker probably was not the master of his cartographic fate when it came to the look of his printed product, especially when the objective involved anything unusual. We cannot now document precisely even a few of what must have been numerous instances when the engraved maps produced by the printer differed markedly from the image in the mind of its cartographic parent. On the other hand, there is some evidence that what started out in the cartographer's mind is not necessarily what the burin produced. For example, some of the engraving of the copperplates for the celebrated *Physikalischer Atlas* of Heinrich Berghaus, prepared in the late 1830s and 1840s, was done in Gotha from compilation worksheets supplied by Berghaus and his compilers in his Berlin Institute. That the Gotha engravers were rather independent is indicated by one of Berghaus's comments in a letter referring to his checking the proofs from Gotha: "Twenty-five years of experience and practice have made me so lenient that I have got used to overlooking [the] many changes the engravers allow themselves."[35]

A rather extreme example of what could happen between the cartographer's image and the engraver's production is provided by a small but unusually important atlas in the history of thematic cartography. In 1806 Carl Ritter, the great German geographer, then a tutor to a wealthy family in Frankfurt, published a little atlas of six maps of Europe that for the first time attempted to portray the interdependence of major geographical phenomena such as surface configuration, climates, plants, and animals.[36] Ritter was a good draftsman, and when he was a student, he had been strongly advised to become a draftsman-engraver. Fate decreed otherwise, however, and instead Carl Ritter became a teacher-scholar of geography, and his name in the history of geographic thought rivals that of Alexander von Humboldt.

Because of his conviction that the configuration of the terrain was of profound importance to an understanding of geographical distributions, Ritter included at the beginning of his little atlas a map entitled, in translation, *The Surface of Europe Depicted as a Bas-Relief by Carl Ritter.* Fortunately, two of his own actual manuscript drawings or worksheets for his *Bas-Relief* have been preserved and are now in the Staatsbibliotek Preussischer Kulturbesitz, Berlin.[37] They are wash drawings in color. The more finished version of the two was probably the map that was given to J. Carl Ausfeld who engraved the other maps for the atlas, although his name is not specifically attached to the *Bas-Relief.* At any rate, the rather attractive watercolor compilation turned out poorly when converted to an engraving.[38] Gerhard Engelman, one of the foremost students of early thematic cartography, put it succinctly when he wrote, "The general impression of these colored original designs surpasses by far the black and white rendering in the atlas."[39] That the engraver may have had trouble with this map is suggested by the fact that when the first printing of the little atlas appeared in 1806, the *Bas-Relief* was absent. Later in the same year a second printing included it. To obtain the effect of continuous tone required a combination of line and aquatint methods of working on the copperplate. Ritter's conception of a shaded plastic relief rendering came just a few years early for the technology. The new process of lithography could have accomplished his objective with relative ease in another decade or two.

The advent of lithography heralded the beginning of a basic shift in the relation between the printer and the cartographer. Diagram 1.3, which is merely a combination of diagrams 1.1 and 1.2, attempts to show the timing as well as the fundamental nature of this change. In the first two periods the woodcut relief and the copper engraving intaglio printing processes both required that a skilled craftsman interpret and render the cartographer's wishes, and this gave to the cartographer-printer team a sort of client-architect relationship. A marked change occurred in the nineteenth century as the tempo of experimentation and innovation stepped up, and by the end of the century a sort of "do-it-yourself" period was developing. To be sure, in the early 1800s the planar printing from stone required skilled craftsmen, many of whom were engravers and lithographers. They still had a strong influence on much of finished, detailed map design. With the perfection of the transfer process in the first half of the century, but especially the incorpora-

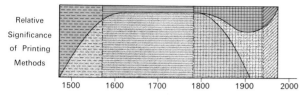

Relative Significance of Printing Methods

1.3 The timing of the periods of printer-cartographer relationships and the utilization of the major printing methods (a combination of diagrams 1.1 and 1.2)

tion of photography in the latter half, the die was cast. With the aid of these techniques anything that could be drawn could be reproduced. The tremendous impact of photography was felt throughout all the printing processes. For example, line and halftone photoengraving, both relief processes, were developed and flourished largely in response to the age-old need to incorporate maps and other illustrations with letterpress text in books. With all these methods, anything anyone drew could be duplicated regardless of its quality. One of the more obvious ways in which this independence is revealed is in the process of lettering the map.

We referred earlier to the ingenious efforts of the printer to incorporate movable type in woodcut maps and that it was not successful in competing with the greater versatility of copper engraving. The desirability of the relief printing plate, either map or pictorial drawing, which could be incorporated with letterpress text, never ceased to spur attempts to develop a workable process. One of the more curious is the attempt of printers to employ, in addition to movable type for the lettering, type cut in the forms of all the other kinds of marks usually employed on maps such as city symbols, mountains, rivers, and so on (fig. 5.1). This was tried as early as the 1770s by Haas, Preuscher, and Breitkopf and was sporadically attempted for more than a century thereafter.[40] As a workable technique it was doomed by its inherent limitations and especially by the newer techniques of cerotyping and transfer lithography.[41] By the third quarter of the nineteenth century movable type could easily be incorporated in a map, and so, after more than three centuries, the cartographer was finally free to make his own choices for lettering.

The incorporation of the photographic process with lithography and the development of techniques for producing relief photoengravings from original artwork allowed the cartographer to bypass the interpretive craftsman entirely.[42] The net result was that increasingly during the latter part of the eighteenth century and into the nineteenth century the cartographer came to be able to do as he pleased since the printer was becoming capable of duplicating anything given him. Although we are a little too close to it to obtain an adequate perspective, it seems likely that we will look back on the first part of the present century as constituting a regressive period in map design.

Several centuries had developed strong conventions in cartographic design, and the corps of engravers did not all expire suddenly with the development of lithography and photography. Their traditions were kept especially strong in the major mapmaking establishments, both the great public national surveys and the private, such as Justus Perthes or W. and A. K. Johnston. But cartographers and engravers are mortal and began to be replaced with what were called cartographic draftsmen, or even the geographer or author who merely thought he had the talent to make maps. Many did not, and accordingly there grew up a class of amateur cartographers, amateur in the sense of not really being qualified, who made maps for books or who became illustrators in commercial offices or government agencies. And there were many of them. The consequences upon map design were profound. This kind of cartographer had no traditions, knew nothing about lettering, and was generally completely untrained graphically. The expert craftsman-engraver who had been a real working partner in the mapmaking process for hundreds of years ceased to exist, and this was serious since he had functioned as a kind of design filter. So far as do-it-yourself cartography was concerned, the printer had simply turned into a duplicator, and neither the newly emancipated cartographer-turned-draftsman nor the printer-turned-copier had the background to carry on the cartographic tradition.

That period of the latter part of the nineteenth century and up to the middle of the twentieth century saw an increasing number of poorly designed maps, at least in the United States. Because duplication was cheap and easy we find these bad maps everywhere, in professional journals, textbooks, brochures, atlases, and so on. In a sense the evolution of the relationship between the printer and the cartographer had come full circle in some four centuries. Five hundred years ago neither the printer nor the cartographer was prepared to approach that very complicated graphic production, the duplicated map, and the same seemed again to be true not so long ago.

Things seem to be changing rather rapidly and dramatically in the latter part of the twentieth century. The printer has become amazingly expert in controlling the camera, the screen, the plate, and the press, while the cartographer is becoming very much more sophisticated about the technical aspects and perceptual intricacies of graphic communication. Printers and cartographers are cooperating closely in the remarkable technological developments associated with modern map construction and processing, involving such innovations as plastics, scribing, proofing systems, stick-up lettering, and so on. Certainly we are now inaugurating a new era in this still essential alliance between the mapmaker and the map printer.

Today, however, the association is quite different from the relationships that have existed during most of the last five hundred years. In the past, one way or another, map design has been strongly influenced by the limitations and capabilities of the printing methods and by the traditions and conventions that grew up because of them. Now practically anything is possible. Furthermore, the distinction between the cartographer and the printer is becoming hard to pinpoint, both in terms of the technical processes they employ and even in the subjects they investigate. We cite two examples.

Recently the printers at the Aerospace Center of the Defense Mapping Agency have developed a system of color identification and have gained its acceptance by all agencies of the DMA. In connection with this they have had to study the perception of tones on maps and in so doing have added to our understanding of the visual gray scale.[43] In the opposite corner we find cartographers proposing methods that up until now, have been largely in the domain of the printers. One of these is technically interesting. Almost magical is a procedure in which one can simply draw single lines and, entirely by photographic manipulation, transform them into double lines.[44] In doing a road map, for example, this could save an enormous amount of time—and frustration—since the scriber need not worry about normal double-line junctures, a perennial headache in map construction. The important point here, however, is that both these innovations have been proposed by persons working in opposite fields, so to speak. On one hand, the printer at the Aerospace Center is investigating perception and, on the other, the cartographer is delving into photographic manipulation. During a considerable share of the last five hundred years comparable professionals probably would have felt they should have minded their own business.

This examination of the changing working relationship between the cartographer and the printer must now be brought to a close. The review has been rather cursory, not only because of the long span of time involved, but also because we actually know relatively little in depth about these important matters. As was pointed out at the beginning, most students of the history of mapping have focused upon the map as a record of the development of geographical knowledge, through exploration and survey, and have paid scant attention to the actual processes by which the acquired data were transformed into the duplicated, graphic map. Except for decorative characteristics, most of the basic elements of cartographic design have been ignored. We badly need the cartographic equivalent of a series of studies in literary history in which we examine the origins and spread of styles, the development of "schools" of cartographic design, and the relation of mapmaking to the other graphic arts. Certainly one of the most influential sets of factors in that transformation we call cartography has been the assemblage of constraints and opportunities provided by methods available to the printer, and, by examining their evaluation, we have tried to emphasize that these technical matters are critical to the graphic character and quality of the final map.

2 | The woodcut technique

David Woodward

"Woodcut" and "wood-engraving": some definitions. The woodcut technique represents the simplest and most direct concept of printing: the transfer of ink from a raised surface to paper using direct vertical pressure. Following Hind, it is convenient to use the term "woodcut" as a generic term to include a whole class of relief prints, i.e., woodcut, wood-engraving, and metalcut, disregarding the material used for the printing surface.[1] But when referring to specific maps, it is necessary to draw a distinction between "woodcut" and "wood-engraving."

The finished print from a woodcut is conceived as black lines on a white ground. To achieve this, the linear design on the woodblock is left in relief, with the non-printing areas carved away with an assortment of knives and chisels. The wood used for such a printing block, preferably medium-grained but easily worked (such as apple or cherry), is commonly cut from the tree as a plank, parallel to the grain of the tree. Blocks for woodcuts are thus occasionally described as being cut "on the plank." It is important to realize that the technique is more akin to carving than engraving, as reflected in the French *tailleur des molles* or the German *Formschneider,* literally "form-cutter."

In contrast, a "wood-engraving" is conceived as white lines on a black ground. Instead of carving the wood with a flat-bladed knife or a chisel, it is engraved by pushing a burin or graver along the surface of the block. The polished cross sections or end-grain of a close-grained wood such as boxwood are particularly suitable for this type of work. Since the blocks are printed as woodcuts, the engraved lines print white and the untouched portions black. Except for small maps in books, very few maps can be described as wood-engravings in the strictest sense: figure 2.1 reproduces one of the rare examples.

A historical sketch. While the cutting of pictures or text on wood for the purpose of block printing dates back to the eighth century A.D. in the Orient, it was apparently not until about A.D. 1400 that the art reached the Western world.[2] The earliest dated European woodcuts that have yet come to light are the Brussels Madonna of 1418 (Schreiber 1108) and the Buxheim Saint Christopher (Schreiber 1349).[3] In its early years, before the invention of movable type, the technique was used for textile

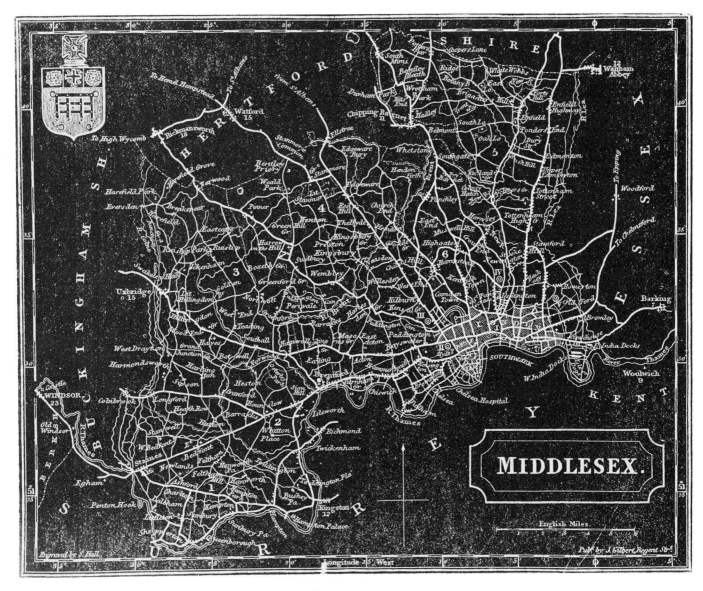

2.1 "Middlesex," in *Guide to Knowledge,* London, James Gilbert, 1832.
Courtesy of the author

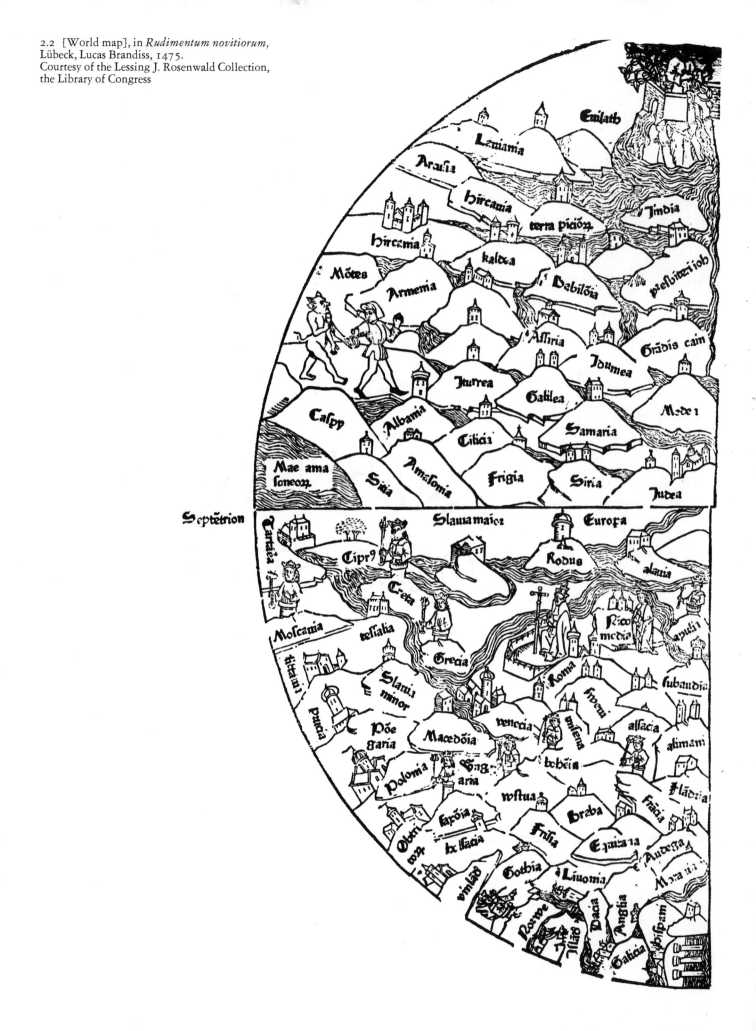

2.2 [World map], in *Rudimentum novitiorum,*
Lübeck, Lucas Brandiss, 1475.
Courtesy of the Lessing J. Rosenwald Collection,
the Library of Congress

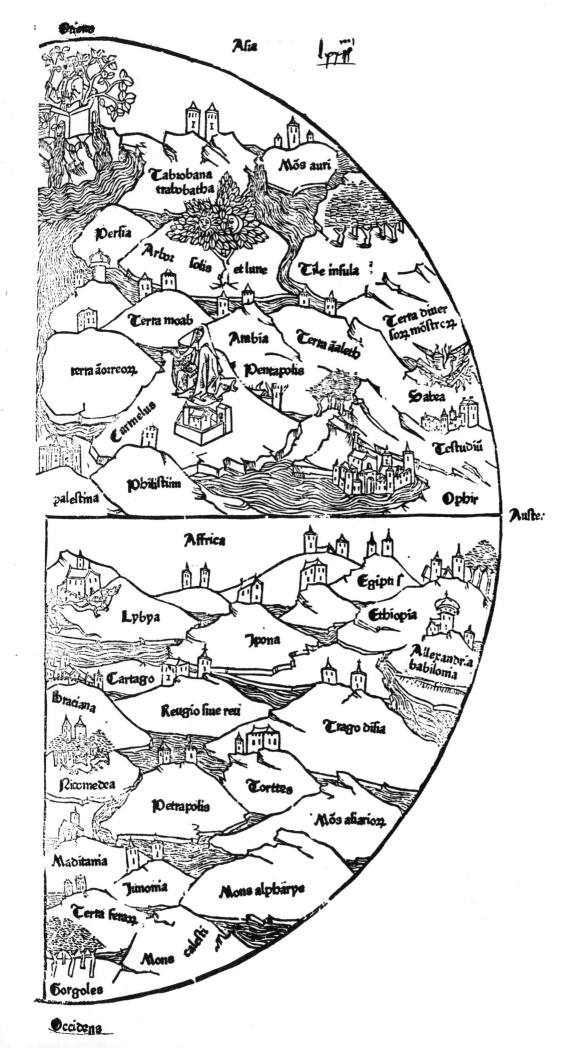

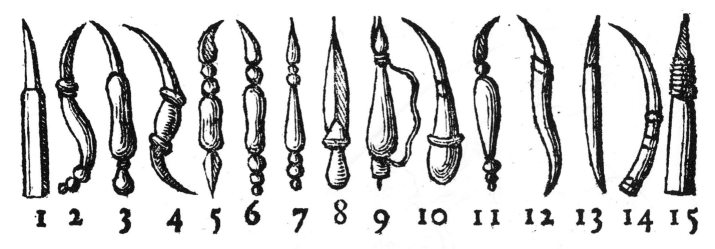

2.3 Woodcutting knives, from J. M. Papillon, *Traité Historique et Pratique de la Gravure en Bois,* Paris, P. G. Simon, 1766, 2:33

2.4 Der Formschneider, in Jost Amman & Hans Sachs, *Eygentliche Beschreibung aller Stände auf Erden,* Frankfurt, S. Feyerabend, 1568

2.5 Part of an original and block used in the production of Philipp Apian's *Bayerische Landtaflen,* Ingolstadt, 1568. sheet 11.
Courtesy of the Bayerisches Nationalmuseum, Munich

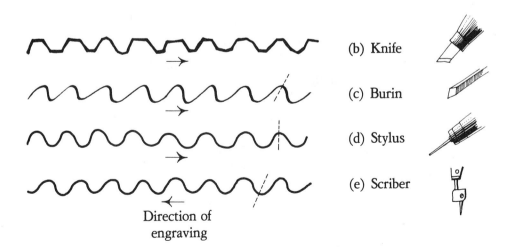

(a)

(b) Knife

(c) Burin

(d) Stylus

(e) Scriber

Direction of engraving

2.6 The effect of four types of engraving tool on line character

2.7 Part of the original wood block used to print *A New Map of England and Wales with ye Roads & Distances of ye Principal Towns in Measur'd Miles from London* [ca. 1800?] Courtesy of the Yale University Library Map Collection

Textur   Fraktur

Schwabacher   Antiqua

2.8 The transition from *Textur* to *Antiqua* typefaces

2.9 G. A. di Vavassore, [Chart of the Eastern Mediterranean], Matteo Pagano, 1539.
Courtesy of the National Maritime Museum, Greenwich

2.10 "Itinera Filiorum Israel ex Aegypto," in Heinrich Bünting, *Itinerarium et chronicon ecclesiasticum totius Sacrae Scripturae*, Madgeburg, A. Duncker, 1957
Courtesy of the Trustees of the British Museum

Neuburg Schongu Saulbach Vetzling Walt münche

Oetting Reichnhall Berchiesgadn Scherding Lech flus

Adeltzhausn Diser gradus jn der Obern München Abach
        bnnd Vntern Leisten/thůn
Emůndt Sechs ain Teůtsche meil. Newkirchen

Beůrbach Deren gradus so auff der Viechtach
        Seitten hieneben Ver,
Veldolfing zaichnet / thůn Vier ain Schmihen
        teůtsche Meil wegs.

Pfarkirchn Pfreimbt Pfaffenhonen Peisserhof Schwabn

Peilstain Passaw Peiserhof Pechtal perg fl. Pierpamm

Nůrmberg Lauff Rot Hiltpoltstain Weissenburg Laus

Herschpruck Vttlhouen Ortnburg Neůmarckt Dacham

Möringen Rotnburg Regen Aldorff Wert Sultzbůrg

Biburg Freistat altn Landaw Isch. Fricting Jlm flus

Schwindeck Waltenreit Minstng Tetnpach fl. Teisnach

Garthausen Schwainperg Riet Euch. Nab Tnessnpach

Oberndorff Am hart Reishersdorff Alling Wamsach

2.11 Impression from an Apian stereotyped plate before cutting, in I. C. F. von Aretin, "Die Stereotypen in Baiern . . ." *Bytragen zur Geschichte und Literatur* 2 (1804): 72

2.12 The reconstruction of the area occupied by type metal in the case of inserted movable type and a stereotyped plate, from *Imago Mundi* 24 (1970):46

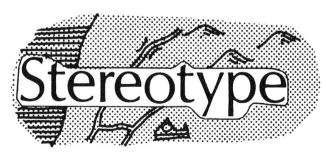

2.13 Comparison of two editions of Sebastian Münster's *Cosmographia*, 1544 and 1564

2.14 Detail from Jacopo de' Barbari, *Venetie*, 1500
Courtesy of the Newberry Library

2.15 Abraham von Werdt, [Representation of a printing office], ca. 1650.
Courtesy of the John M. Wing Foundation, the Newberry Library

2.16 Detail from the world map on a cordiform projection in
Claudius Ptolemaeus, *Liber geographiae.* Venice, J. Pentius, 1511.
Courtesy of the Trustees of the British Museum

prints, playing cards, religious prints or broadsides, and probably block-books. It was not until about twenty years after the introduction of movable type that books started to appear with subject illustrations; the earliest known is *Der Ackermann aus Böhmen,* printed by Albrecht Pfister of Bamberg about 1460. It has been estimated that about one-third of all fifteenth-century printed books were illustrated, and the majority of these illustrations were woodcuts.[4]

The first known printed map in the Western world appeared in a 1472 printed edition of the dictionary of Saint Isidore of Seville.[5] It was a simple woodcut T-in-O map of the world in the form that had been in use since Roman times (fig. 1.1) and thus achieves prominence less from its content than as its status as a "famous first." It heralds no printing or cartographic innovations, and probably deserves less attention than it has received. In contrast, the maps in the *Rudimentum novitiorum* have not been studied sufficiently considering their importance (fig. 2.2).[6] These maps are the earliest printed maps not derived from a classical source and exhibit many cartographic and historical facets worthy of study.

While it may be surprising to some that printed maps did not appear until thirty years after the invention of movable type, a similar time lag was experienced by other classes of technical illustration. Thus, the first printed illustrations for a technical book are not found until 1473 in *De re militari,*[7] and the earliest printed music also did not appear until the early 1470s.[8] The reason for this time lag would repay further study: it may have been a small demand for illustrated technical printed books until this period or the lack of suitable engravers and know-how.

The use of the woodcut technique for maps flourished from the 1470s until the middle of the sixteenth century. While many examples of woodcut maps emanate from regions other than central Europe during this period (for example, Paris and Venice), the main scenario of woodcut map reproduction was centered north of the Alps, in the Rhine Valley, Bavaria, and Swabia.[9] Here there was already a strong tradition of wood carving and a correspondingly good supply of craftsmen. The momentum of such a tradition helped to carry it through the first half of the sixteenth century. Further, most maps of the incunable period were intended to be printed in books, and the woodblock was well suited to this use, as it could be placed in the form with the type and both could be printed at once. The woodcut could be printed on a simple screw press as used by Gutenberg, which was in common use by printers of the period and would have been readily available for the printing of maps. The rolling press, needed for printing from copperplates, was not generally available north of the Alps until later in the sixteenth century.

The rise of the Italian and, later, the Flemish map publishing centers in the mid-sixteenth century effectively superseded the woodcut. Based on the intaglio technique of copper engraving, the new breeds of cartographers found a versatility and fineness of line in this method of engraving with which the woodcut could not compete. As the style of woodcut in general gradually deteriorated toward the end of the sixteenth century, woodcut maps became increasingly sparse. Copper engraving, of course, had been used for maps as early as 1477 for the Bologna Ptolemy, and the quality of the engraving of the 1478 Ptolemy illustrates a very early expertise in the technique. The woodcut and the copperplate thus existed side by side for the first century of map printing. Copper engraving was favored south of the Alps; woodcut to the north.

The seventeenth and eighteenth centuries yielded very few woodcut maps. Most of those that exist were later impressions of sixteenth-century work[10] or prepared for book illustrations in an ad hoc manner,[11] using materials and equipment readily available in the common printing office. When compared to the volume and quality of maps produced by copper engraving during these two centuries, these woodcut maps take on the character of curiosities. But the use of the woodcut during this period does illustrate the point that, with the barest minimum of equipment available, the woodcut was the technique of choice.

In the nineteenth century, a strange hybrid of woodcut and wood-engraving was used to produce thousands of small and usually substantively insignificant maps in textbooks, encyclopedias, and newspapers. These maps were woodcuts in the sense that they were conceived as black lines on a white ground (in contrast to the map of Middlesex in fig. 2.1), yet were wood-engravings in that they were engraved on the end-grain of boxwood with a burin. The appearance of such a painstaking and awkward method of reproducing maps bears witness to the severe need in the first half of the nineteenth century for a graphic relief printing process that could produce blocks for the power press. This need was partly met by the development of wax-engraving in the eighteen-sixties and 'seventies and, later, by the photomechanical line block in the 'eighties and 'nineties.

Source materials. Our knowledge of the materials, tools, and techniques associated with the woodcut is derived from two main sources, writings and artifacts. Since the art has been practiced in some form or another continuously until the present, we would expect some aspects of the technique to have been handed down through the years with little change. Thus, while we do not have an extant technical manual devoted exclusively to woodcut and wood-engraving until Papillon's treatise in the eighteenth century,[12] the methods as he described them are possibly similar to those used in earlier centuries. But Papillon's historical descriptions are not always accurate and should be used in conjunction with other sources. The lack of earlier descriptions of the technique precludes a more definite application of Papillon's book to an earlier time. Later manuals tend to emphasize the technique of wood-engraving, as distinct from woodcut. The reasons for this were various. Wood-engraving involved new skills and did not have the long craft tradition of woodcut. Furthermore, wood-engraving became a craft industry in the nineteenth century, and technical manuals were in demand for apprentices. Whatever the reasons for the emphasis, information on the older woodcut technique is scarce in such works: that which does exist is usually derived from earlier manuals.

The best summary of the literature of the woodcut technique in English is still the first chapter of Arthur M. Hind's *An Introduction to the History of Woodcut.*[13] As in most surveys of graphic arts media, Hind's work leans more to the artistic than to the craft aspects of the technique; many technical questions are left unanswered. In addition, he does not—and cannot be expected to—include the special technical problems that maps posed the woodcutter.

In the absence of major written sources of information regarding the woodcut technique, we are obliged to examine the available artifacts, woodblocks and impressions from woodblocks, in order to fill in certain details. Original woodblocks used to print maps are scarce. The following list gives a few examples:

Date	Map	Location
1500	Jacopo de' Barbari, *Venetie*	Museo Correr, Venice
1515	Johannes Stabius, [Eastern hemisphere]. Reprinted from the original blocks, 1781	Graphische Sammlung, Albertina Bibliothek, Vienna
ca 1560	Hadji Ahmed, [World map]. Not printed until 1794	Biblioteca Nazionale di S. Marco, Venice
1566	Jodocus Murer, [Map of the Canton of Zurich]	Staatsarchiv des Kantons Zurich
1568	Philipp Apian, *Bayerische Landtaflen*	Bayerisches Nationalmuseum, Munich
eighteenth century?	Cartographer unknown, *A New Map of England and Wales . . .*	Yale University Map Collection, New Haven, Connecticut

These blocks provide evidence of the size, thickness, and type of wood used, the tools employed, and, in the case of the Murer and Apian blocks, positive evidence of the use of sterotyping, to which I shall allude later.

A much more abundant, if not always as eloquent, source of technical information is a study of the actual impressions from the blocks, the printed maps themselves. One of the surprisingly undeveloped skills of historians of cartography, as evidenced by the literature, is that of accurate, precise, and systematic descriptions of the printed map as a physical object. Yet this, as in all other empirical fields of study, is the very basis of the discipline. In the case of the woodcut technique, where written manuals and original map woodblocks are scarce, the map itself may be the only evidence available for certain aspects of the technique. For this reason, this essay makes frequent use of original impressions as sources of information.

FASHIONING THE WOODBLOCK

As we have seen, a distinction is usually drawn between "woodcut" and "wood-engraving" and most of the maps printed from woodblocks in the fifteenth and sixteenth centuries (the main period under consideration in this essay) fall in the "woodcut" category. It is therefore more accurate to describe the preparation of the printing surface as "fashioning" or "cutting" than as "engraving."

Materials. The wood for the block was characteristically sawn parallel to the grain of the tree and planed down to produce a plank. Figures for the maximum size of such blocks are not available; it is often difficult to perceive from the impressions whether or not a composite block has been used. Generally speaking, the larger the block, the greater the tendency to warp, and prevention of warping of the printing surface was critical, for obvious reasons. Thus, for large work, several small blocks were often butt-jointed or mortised together. The straight white lines often appearing on woodcut impressions indicate clearly where such joints have loosened.

The blocks were planed to approximately ⅞″ so that they could be printed with movable type. Since there was no accepted standard type-height in the fifteenth and sixteenth centuries, the woodblocks must have required considerable make-ready to bring them to the height of type. This was presumably done by pasting pieces of paper to the underside of the blocks, as is the practice in letterpress printing today.

Hind summarizes the types of wood used for woodblocks as of beech, sycamore, apple, pear, or cherry.[14] These are medium-grained materials able to hold fairly fine detail, yet they are easily worked with a flat-bladed knife. Boxwood, often erroneously thought to have been introduced as material for woodblocks by Thomas Bewick at the close of the eighteeenth century, was actually used as early as the fifteenth century. This is indicated not only by some particularly detailed work requiring close-grained wood of this type that dates from that century, but also by a possible reference to the use of the material for wood type in a colophon to one of the books printed by Johannes Trechsel in the fifteenth century.[15] To my knowledge, the composition of extant woodblocks for maps has not been systematically analyzed.

Transferring the design. In woodcut, three main methods are recognized of transferring the design in the artist's mind to the surface of the woodblock: (a) by drawing directly on the block, (b) by pasting a drawing on the block and cutting through it, and (c) by transferring a drawing to the block in "carbon-paper" fashion.[16] Here should be emphasized an important distinction between general pictorial work and cartography. It is generally easier to reverse a sketch than a map, because the latter almost always has some lettering on it. One would consequently expect the author of the drawing (whom we shall loosely term the cartographer) to choose a method of transferring his drawing without having to reverse it. This would be by using method (b): the drawing could be varnished on the back to bring the image through, pasted face down on the block, and cut through from the back. This, of course, would destroy the cartographer's fair drawing, which might account for the scarcity of manuscript maps actually used in cutting a woodblock. On the other hand, there is some evidence, for example, the reversed *Ns* in the Ulm Ptolemy of 1482, that the draftsman indeed had to face the problem of reversing the image on the block. We would also expect a natural inclination on the part of the cartographer to preserve the original fair drawing, in which case method (c) would be more appropriate. In this method, chalk was applied to the back of the manuscript map, which was laid face up on the block. The detail would then be transferred to the block by tracing over the manuscript with a hard stylus. For this method, the manuscript map would have to be drawn in reverse. Since no examples of such documents survive, to my knowledge, and no mention of the method specifically for cartography apparently exists in the literature, proof is not forthcoming that the method was actually used for maps. Indeed, it is wise not to become unduly specific in ascribing any one method as dominant, for they were probably all used at every stage of the woodcut map's history by one craftsman or another.

Tools. The traditional tool of the *Formschneider,* which appears with the monograms of several woodcutters, was the flat-bladed knife, not unlike the modern Xacto or craft knife (fig. 2.3). According to Papillon, the blades of such tools were cut from extremely hard metal such as sections of clock springs and set in thin wooden handles that would fit in the hand like a pen.[17] Very careful attention was taken in the sharpening of such tools: a dull knife could make cutting impossible for even the most skilled craftsman.[18] It is significant that Jost Amman's woodcut of the *Form-*

schneider at work features prominently on his bench a sharpening stone, ready for use after every two or three cuts with the knife (fig. 2.4).

Contrary to common opinion in the literature, the flat-bladed knife was not the only tool used by the *Formschneider,* and the usual distinction between woodcut and wood-engraving based on the type of tool used (knife versus graver) is not altogether a happy one. The existing blocks of Apian's map of Bavaria show quite clearly the marks of gouges and chisels in addition to the traditional knife (fig. 2.5). In addition, the detailed parallel line-work perhaps shows evidence of the use of the graver, and if Hind is correct in saying that "work with the graver cannot be done on the plank,"[19] such use implies that these blocks were cut on the end-grain. This is a difficult question that must be deferred until these blocks are examined systematically. It is nevertheless clear that the early craftsman used a wide variety of tools and was by no means confined to using the simple knife.

Line-work. The tools were probably used for line-work in the following manner. The flat-bladed knife would be used to outline the design on the block and the wood on either side of the line removed with gouges and chisels. Since wood is a resistant medium, it places certain restrictions on the craftsman, and these are often reflected in a characteristic woodcut style. Figure 2.6 summarizes the character of lines produced with four fundamentally different engraving tools. We will assume (a) to be the model—perfectly smooth and regular.

The woodcutter's flat-bladed knife will tend to produce a characteristically angular line, often of uneven thickness, as in (b). The knife is difficult to control against the resistant medium, resulting in frequent slips and irregularities. On the other hand, the burin or graver used by copper-, steel-, or wood-engravers, held firm against the palm of the hand and pushed away from the operator, is capable of a flowing, elegant, delicate line unmatched by any other engraving tool. But, as a result of this pushing motion, it is natural both to "overshoot" and engrave more deeply in one direction than another, (c), resulting in a slight skewness and a thick- and-thin character.

A pointed tool such as the etcher's stylus produces characteristically rounded curves, (d), that probably approximate the character of (a) more closely than the other tools described here. Held in the hand like a pencil, it allows considerable control, especially when engraving in material of little resistance, such as etching ground. Since the cutting part of the tool is round, the direction of the line has little bearing on the resulting thickness.

The swivel scriber represents the most modern of the tools described here. In this tool, the cutting edge is automatically maintained at 90 degrees to the line being engraved, because the point is allowed to swivel freely as it is pulled toward the operator. There is, however, a short lag that often results from this pulling motion; in other words, the tool may not change direction precisely when the operator requires it, but slightly afterward. The experienced scribing draftsman obviously compensates for this, but there is a tendency for the tool to behave this way.

Similar clues may be perceived by examining the ends of the engraved lines. In the copperplate engraving, the line tapers off, usually to a point; in etching, the end of the line is usually rounded; while in woodcut it is square—this results from the nature of the tools and techniques used.

Point symbols. Some difficulty was encountered in the cutting of regular point symbols such as circles and dots in wood. The medium was simply not well adapted to it. Dots intended to be circular turn out to be square; circles intended to be round tend to become angular. One technique used to alleviate the problem of cutting dotted lines in wood was to drive pins into the block, as revealed by the Yale woodblock (fig. 2.7).

Area symbols and tone. The production of tone on woodcut maps presented a difficult problem. It was laborious to cut a dotted or line pattern, and when done, this tended to hide other detail underneath. Most difficult of all was cross-hatching, where tiny lozenges of wood had to be excavated between the lines, as in some of the work of Dürer. The end result was that woodcuts often avoid the use of tone altogether, presenting an open, stark appearance.

Lettering. From the earliest woodcut to modern techniques, lettering has been a pervasive problem in cartography. Almost all maps contain some lettering, and the difficulty arises in combining the lettering with the line-work when reproducing the map. This is a problem shared by music printing, and indeed the developments of map and music printing have interesting parallels that invite study.

The most obvious way of lettering a woodcut map was to cut the letters directly on the block by hand with the other detail. The cutting of lettering encountered the same problem as cutting of lines: square and straight lines were easier to cut than rounded lines, and the same crude and primitive character resulted. Hence, the lettering style known as *Textur* was particularly well-suited to the woodcutter's requirements. Textur is a square, severe, and simple style that lent itself well to the flat-bladed knife. It was among the best of the formal book hands of the late Medieval period, and formed the basis of Gutenberg's designs for the type of his 42-line Bible. Later, when the influence of the softer Renaissance styles spread through central Europe, the straight-shanked character of Textur was modified to the rounder character of Schwabacher and Fraktur on which most of the sixteenth-century German type styles were based (fig. 2.8).[20] The rounder the styles became, the more the woodcutter's problems increased, and, when the roman type styles became fashionable, his difficulty had reached its peak.

Handcut lettering on wood shares many of the characteristics of hand-lettering at large. It is versatile in that it can be cut in any size and style, and can be compressed in the space available. Such advantages are not shared by letterpress type. On the other hand, some inconsistency is always present: inconsistencies between letter shape, size, spacing, and verticality. These lend a certain look to the map and announce it as having been hand-lettered. Such inconsistencies were magnified by the resistant character of the woodcut medium.

An excellent example of handcut lettering is seen in the work of Giovanni Vavassore, a Venetian printer, publisher, and woodcutter working in the mid-sixteenth century (fig. 2.9). Vavassore's lettering is skillfully rounded, but exhibits the usual inconsistencies characteristic of woodcut lettering. The letter size and shape is often inconsistent, and the letters frequently stray unintentionally from the vertical. There is evidence that he used rounded gouges or circular punches to form the counters of letters such as "O"; their remarkable consistency probably rules out the use of a flat-bladed knife. Vavassore has made a conscious attempt to follow the rounded roman style, but has omitted or truncated the serifs. He has usually also trimmed the top and bottom of the names to finish them neatly.

The obvious way out of the difficulties faced by the woodcutter in cutting letters by hand was to set them in type. Type had been used with woodblocks on the same page in books before the appearance of printed maps, and the same idea could easily be applied to setting captions or marginal information around the map.

The written sources on this matter are largely in the nature of passing references: most authors, when describing maps on which the technique occurs, have failed to mention the fact. Breitkopf was apparently the first writer to recognize the method, as used in the Venice Ptolemy of 1511.[21] References to the technique on these and other maps are also found in Chatto (1861), Winsor (1884), Stevens (1908), Huck (1910), Stevens (1928), Anesi (1948), de Dainville (1964), and Skelton (1969).[22] But there were technical difficulties in using type on the face of the map itself. The most direct method was to chisel a slot in the block and insert the type; this was probably done for a large number of late fifteenth- and early sixteenth-century woodcut maps, starting with the two maps in the *Rudimentum novitiorum*. For the setting of cartouches and legends, which frequently required much setting of small type in one or two slots, the method must have proved a time-saver. But the task of positioning and cutting one slot for each name on the map was more demanding. In addition, as Stevens suggests, the block might become so honeycombed with holes that it would warp or split.[23]

Efforts to circumvent these problems were numerous. In the first place, many maps show evidence of having used two impressions, one for the type and one for the handcut detail. Examples are found in the world map by Sebastian Münster in the Basel 1532 edition of Grynaeus's *Novus orbis*[24] and the 1597 edition of Bünting's *Itinerarium Sacrae Scripturae*.[25] Such cases are easy to identify as the type invariably overlaps the woodcut detail at some point or another. In addition, since the type was locked up in a separate form, vertical shifting of the type sometimes occurred, as in figure 2.10.

Another method to avoid the labor of cutting separate slots for type, much later in the history of woodcut maps, was a product of Victorian ingenuity: a slot-cutting machine for the type in the wood-engraved maps of the *Penny Magazine,* a popular illustrated journal published between 1832 and 1845. The holes were "pierced with the greatest rapidity by gouges of different sizes acting vertically—put in motion by machinery contrived by Mr. Edward Cowper."[26]

A third method, much more ingenious for its time, was to use a form of stereotyping. Stereotyping is the production of a solid printing plate cast from a mold created by any printing surface (which in this context is movable type). We have fairly precise information on this procedure for woodcut maps from the extant blocks of Philipp Apian's *Bayerische Landtaflen* (1568) and a reference to an original stereotype plate apparently for use with the Apian blocks. Sixty or a hundred names were set up in type, and a matrix was obtained in damp paper or some other molding material.[27] McMurtrie is of the opinion that the impression was taken in sand, a method "known to have been used at an earlier date for the duplication of woodcut engravings."[28] A molten tin or lead composition, probably similar to type metal, was poured into the mold.[29] The result was one of the plates described by Aretin in 1804 and illustrated in figure 2.11. The existence of such a plate in Aretin's time probably indicates that duplicate plates were produced when the map blocks were made, perhaps intended to provide replacements for names that might become damaged or lost.

The plates were cut up, probably with metal shears, and affixed to the woodblock with cement or putty that Hupp describes as a red, gray, yellow, or white mastic.[30] As one of the original woodblocks shows (fig. 2.5), the plates lie in the hollowed-out portions of the block between the lines, interrupting them only infrequently. It is difficult to imagine how these plates were affixed to produce the remarkably even impressions of the Apian map. One method would have been to mount the names on patches of slow-drying adhesive so that they were slightly higher than the woodcut detail, and then to level the printing surface with a flat block, thus pushing the stereotype plates into the adhesive.

The recognition of stereotyping from the map impressions above is a difficult task, as the intrinsic characteristics of stereotypes follow those of the printing type from which they were cast. Hence, the literature frequently contains references to the use of type on maps when it would have been more precise to specify stereotyping.[31] But there are some characteristics of stereotyping that allow an identification of the techniques from the impressions themselves. The first type of evidence involves the relationship between the type and the surrounding detail. When movable type is inserted in slots cut through a woodblock, the woodcut detail will not be found closer to the type than the type body will allow, as in figure 2.12. On the other hand, a stereotyped plate could be trimmed quite easily around the face of the type, as the original Apian woodblock demonstrates. The woodcut detail may thus encroach on the area normally occupied by the type body (fig. 2.12). On the basis of this evidence, the author has demonstrated that stereotyping was probably used on a world map of Peter Apian dated 1530.[32] Using similar evidence, it may be found that earlier maps used the same technique.

Where two or more states of the block exist, the stereotyping technique may often be identified positively. The relative position of letters cast on a stereotyped plate will not move throughout its printing life. Yet the plate itself may drop off or shift between states, as the work of Hupp and Horn shows.[33] For example, in figure 2.13, the word "Gangem" has been replaced upside down sometime between the 1544 and 1564 editions of Sebastian Münster's *Cosmographia*. If the word had been set in movable type inserted right through the block, a new slot would have been necessary to accommodate the new type. But it is unlikely that the printer, in replacing the word, would have cut another slot for the type when one already existed. The obvious explanation is that the name was cast on a stereotype plate, that it became loose during or after the 1544 edition, and was replaced (probably hurriedly) upside down.

While this is conclusive enough evidence of stereotyping in itself, it is supported in a letter written by Münster to Joachim Vadianus in 1538.[34] Part of the letter has been translated by Chatto thus:

I would have sent you an impression of one of the Swiss maps which I have had printed here, if Froschover had not informed me of his having sent you one from Zurich. If this mode of printing should succeed tolerably well, and when we shall have acquired *a certain art of casting whole words* [my italics], Henri Petri, Michael Isengrin, and I have thought of printing Ptolemy's Cosmography . . .[35]

Chatto describes this "mode of printing" as an attempt at logography. Logography, or logotyping, was a method of casting words or groups of letters in one piece, which was invented by François Barletti de Saint Paul around 1776.[36] Logotypes were used in a way similar to foundry type, i.e., they were composed, printed from, and distributed to the type case for future use. On the other hand, stereotypes were cast as thin plates to meet a specific need, and thus had to be melted down and recast to be used again. Chatto's translation of the Latin "dictiones" in the Münster letter as "words" reflects his view that the new method of printing described was logography. Yet "dictiones" may also be translated as "phrases" or "sentences," and we might have expected Münster to use the more common "verbum" had he intended the specific meaning of "word." Thus, if Münster was indeed referring to phrases or sentences cast as one piece, this implies stereotyping rather than logography.

The picture begins to emerge of stereotyping as a technique that was well-known to both Apians, father and son, Sebastian Münster, and probably most of the German and Swiss printers and cartographers of the middle sixteenth century who were still using the woodcut as a method of map reproduction. It was certainly known to Joos Murer, whose map of the Canton of Zurich (1566) was printed from blocks using stereotype plates. These blocks are preserved in the Staatsarchiv, Zurich. Confirmation of other specific uses of the technique needs further research.

Corrections and revisions. Since maps quickly become out-of-date, the ability to make corrections and revisions has always been the concern of the cartographer and printer when choosing a method of reproduction. The problem was much less acute in the fifteenth and sixteenth centuries than it is now, so that the difficulty in making corrections in a woodblock was probably of less concern then than it would have been in later centuries. But the problem was there, nevertheless, and was, no doubt, another reason for the supersession of the woodcut by copper engraving, in which corrections and revisions were technically much easier.

Corrections to any relief process are difficult. If the original printing surface is marred or cut incorrectly, it can only be restored by cutting out the part to be corrected, inserting a plug of wood, and recutting the detail. Such corrections can usually be recognized by a slight gap between the plug and the original block, expressed as a thin white line in the impression.

The history of one of the blocks of Jacopo de' Barbari's view of Venice (1500) provides an example of a corrected woodblock. The original block bore the date 1500 and a representation of the Tower of Campanile, correct for that date. After 1513, when the tower had been rebuilt, the view was reissued with the changes. Plugs were skillfully inserted in the area, and the new representation cut (fig. 2.14). Probably at the same time, the date 1500 was cut out. A still later state of the block has the date reinserted.

Corrections could be made more easily in woodblocks with inserted type or stereotype plates. Different states of the block are consequently particularly common in maps printed by this method. Replacement of type could be done for several reasons, for the purpose of making corrections, or to renew old, worn, or lost characters. In some cases, for example in Münster's *Cosmographia,* the type on the blocks has been varied to suit the language and taste of the reader, with Roman type for the Latin editions, and Schwabacher or Fraktur for the German editions. Type may also have been used to correct badly or wrongly cut lettering. The existence of a few names on a map set in type with the rest hand-cut is an indication that this has been done. Examples are seen in the Strasbourg Ptolemy of 1520 (British Isles) and the Boston variant of John Foster's *A Map of New-England* (1677).[37]

Corrections could be made without changing the woodblock. In some copies of the Venice Ptolemy of 1511, it appears that some detail has been overprinted with a second block to make the required additions. Another method, exemplified in the same work, was to paste on small pieces of paper to reverse the locations of two place names.

PRINTING THE BLOCKS

Presswork. Since both the woodcut and metal type were relief printing methods transferring ink from the raised portion of the printing surface, the woodcut could use the common printing press developed in the middle of the fifteenth century and still in use for hand-printing. This was a great advantage, not only because such presses were readily available, but also because a woodblock could be printed with a page of type in one impression.

The method of printing woodblocks was essentially the same as printing from type, a procedure that has been described and illustrated many times in the literature,[38] so that only a brief summary is necessary here (fig. 2.15).

The block was placed face up on the chase and inked with leather dabbers. The ink used in woodcut printing was thicker and stickier than in intaglio printing, since it had to stay on the raised printing surface. A sheet of handmade paper was dampened and laid on the tympan, held and positioned by two small points. Hinged to the tympan was a frame, the frisket, covered with frisket paper, which was cut out to mask out the margins of the impression during printing, so preventing stray ink from marring it. The frisket was folded on top of the tympan and both were folded over the woodblock in the chase. All were then slid along the bed of the press under the platen and subjected to vertical pressure with a handle and screw. The procedure was reversed to remove the print.

Variations in the design of the press were only slight until Stanhope's iron press of 1800. Before this, the great vertical stress involved during printing required heavy braces to be constructed between the press and the ceiling of the printing office, as illustrated in the contemporary woodcuts of Amman, Dürer, and others.[39]

Color printing. The great majority of woodcut maps were printed with black ink in one impression. Examples of color printing that exist were therefore probably in the nature of experiments that apparently proved negative. Color printing had been used for initials and illustrations several times during the fifteenth century: it was not until the Venice Ptolemy of 1511, with its use of red and black ink for different classes of geographical names, that the first use for maps appears. On the basis of a single, but not insubstantial, piece of evidence, the method of printing the Venice Ptolemy in two colors may be reconstructed. In figure 2.16 the word "come" is printed in *both* red and black, the black overprinting the red. From this, we may observe: (1) that at least two impressions were involved; (2) that the same type was used for both black and red impressions; (3) that the word "come" was accidentally inked red; and (4) that, considering sixteenth-century technology, the registration is good. The tentative conclusion is that the type-inserted block was laid in the press, and the red names only were inked, "come" by mistake (it is split up in groups of two letters similar to other provincial names printed in red). Without moving the block, it was either wiped very clean or the red characters were removed from the block. By this time, the error of printing "come" in red had been noticed, so the type was cleaned and left in the block. The whole block was then inked black, and the printed sheets were again laid in the press for the black impressions.

The other oft-quoted example of woodcut color-printing for maps is the map of Lotharingia found in the Strasbourg 1513 edition of Ptolemy.[40] It is printed in three colors, which vary in hue from one copy to another (*frontispiece* and fig. 1.4). The registration also varies widely. Nevertheless, it is of great interest in the history of map printing, because it illustrates that the desire for color in map reproduction was evident early. The technical ability did not match the idea, however, and I am aware of no other instances of multicolor map printing in woodcut until the work of Firmin Didot and Charles Knight in the nineteenth century.[41] In this field, however, new discoveries are always around the corner, and it is very unwise to be absolute in one's statements about the existence of examples of a particular technique.

It is not unusual to find woodcut maps colored by hand. As with woodcuts at large, it is probable that certain characteristic colors were used at different publishing centers. Hind has attempted a table of the pigments used and their probable local use, based on work by Schreiber and Laurie.[42] The extent to which such generalizations can be applied to the coloring of maps will not be clear until a systematic study of color on woodcut maps is completed.

Style. There is one general aspect of research into the woodcut technique that we have not discussed: the style characteristics of the map as a whole, its general "feeling" and look. For example, how does a woodcut map cut by Johann Grüninger differ from one cut by Jost Amman? We literally have nothing in the way of systematic tools that list the characteristics of each engraver in an ordered manner. The least we can do in the absence of such tools is to build a fund of experience by looking at as many maps as possible and evolving a "feeling" about a particular period or engraving style, much in the same fashion as does a connoisseur of prints or glass paperweights. This of course is not very helpful to the layman, and even less helpful to posterity: the expertise dies with the expert. Thus, we need to develop a vocabulary of words that describes these characteristics accurately and attempt to compile graphic encyclopedias of style characteristics.

Franz Grenacher has demonstrated how the different styles of woodcutting can be used to trace the work of a single craftsman. Based on the style of the "Tabula Nova Heremi Helvetiorum" in the Strasbourg Ptolemy of 1513, which he ascribes to a specific school of woodcutters in Strasbourg identified by Paul Kristeller, he has followed his trail from Strasbourg to Krakow, and perhaps to the Netherlands.[43]

Once we can identify map engravers by the characteristics found on their maps, we are one step nearer identifying the date and place of publication of a map, or even establishing its very authenticity. At the moment, this process requires a veritable research project on each map itself, a fact all too well known by people who study early maps.

3 | Copperplate printing

Coolie Verner

The process of metal engraving was known and practiced in antiquity by goldsmiths in decorating the objects they produced. In order to preserve a copy of a design engraved on an object, the incised lines were filled with lampblack and the design then transferred to paper.[1] In due course, this simple expedient was adapted to printing graphic material on paper; the earliest known dated print is one of a series in the *Passion of Christ* printed in 1446 in Berlin.[2] The application of this process to the production of maps occurred first in 1477 when the twenty-six maps in the Bologna edition of Ptolemy were printed from engraved copperplates.[3] This event inaugurated a new era with far-reaching consequences in the development and diffusion of geographical knowledge.

Before the introduction of printing, maps were limited to hand-drawn copies, which severely restricted the diffusion of information. Such manuscript maps allowed the cartographer great flexibility in presenting geographical data, but they were subject to error in reproduction. Woodblock printing tended to reduce errors of reproduction and increased the spread of information, but the nature of the material imposed severe restrictions on the amount and kind of data that the cartographer could present.

Copperplate printing reduced the limitations imposed by manuscript copies or woodblock printing and enhanced the utility of the map for the graphic presentation of geographical information. Among other things:

1. Maps printed from metal plates could be larger than those from woodblocks. This provided space for the cartographer to include more detailed geographical data.

2. The flexibility and fineness of an incised line encouraged greater precision in the representation of details so that maps became more precise.

3. The relative ease with which metal plates could be altered accelerated the introduction of new or amended data by eliminating the need to prepare a completely new map.

4. Metal plates had a longer useful printing life, which resulted in the reduced cost of a single copy; maps became more nearly within the reach of everyman.

5. Because the act of preparing and printing maps from metal plates required specialization, the production and use of maps was divorced from letterpress,

consequently their appeal was not limited only to that segment of the population that was literate. As a result, geographical knowledge was more widely diffused among all segments of the population.

Along with specialization came the establishment of efficient procedures for the preparation and printing of metal plates. Although precise knowledge of these procedures is scanty, existing evidence spanning several centuries suggests that there were few changes over time.[4]

ENGRAVING

Printing from metal plates is an intaglio process in contrast to the relief process of woodblocks. Intaglio printing from metal plates was accomplished through line engraving, drypoint, etching, mezzotint, or similar methods of incising lines in a plate. The plate used for printing could be any relatively soft metal including gold, silver, iron, zinc, or pewter. For the production of maps, copper has been the preferred metal and line engraving the most useful intaglio method (fig. 3.1).

Preparing the plate. The selection and preparation of a copperplate was a complex and laborious procedure. Faithorne noted: "That Copper is best which is free from flaws, and not too hard, which you may perceive by its yellowish colour, almost like brasse; if it be too soft, you may somewhat perceiv it by its too much pliableness in bending."[5] The plate ". . . must be fully as thick as an half-crown . . ." and it must be planished with a hammer to insure that it is firm and free from holes or flaws. The smoothest side of the plate had to be polished to a mirror-smooth finish—first with a piece of grinding stone and water, which removed the marks of the hammer; next with pumice stone to remove any marks left by the grinding stone, followed by a ". . . fine smooth hone and water . . ." to remove the marks of the pumice, and then the small strokes of the hone are polished away by rubbing the plate with hardened charcoal. The final step in the preparation of the plate involved rubbing it with a steel burnisher until it had a mirror finish. Only then was it ready for use.

The skill and care exercised in the preparation of the copperplate affected the appearance of the impressions pulled from it. The maps appearing in Morse's *Geography . . .*, published in 1789, were engraved by Amos Doolittle and drew criticism from Ebenezer Hazard, who wrote Morse on 27 July 1793 that "Your maps, however, are badly executed, & the plates have not been well polished, hence the prints have a dusty appearance."[6]

Map plates varied in size according to the use for which the map was intended. A plate in the British Museum made about 1690, with a map of the City of Corke, measures 20.0 x 28.0 cm (8 x 11 in) and weighs eleven pounds, twelve ounces.[7] In the nineteenth century the plates for the six-inch maps of the Ordnance Survey were 61.0 x 91.5 cm (24 x 36 in) and weighed thirty-five pounds. A chart produced by the Hydrographic Department at Taunton is 71.0 x 104.0 cm (28 x 41 in) and weighs forty pounds. These examples do not permit any generalizations about the relationship between size and weight.

At first, map plates were small, but the advantage of size quickly became obvious so that plates got larger and larger. Aaron Arrowsmith was noted for the large plates he produced which necessitated special paper that he had made by *"Fourdrinier's* Patent Machinery, thus avoiding the numerous junctions, which too often injure the appearance of good Maps, when printed on many Sheets."[8]

Transferring the design. After the plate had been prepared it was heated and rubbed with a white "virgin-wax" that was spread evenly over the plate by stroking it with a feather. When this had hardened the plate was ready to receive the design.

The material to be engraved on the plate had to be drawn on the white wax in reverse. This was done in several ways. Faithorne advised that the drawing was to be laid face down on the plate and the back rubbed with a burnisher or traced with a pencil that transferred the design to the wax coating.

An alternate method of transfer recommended by Bosse[9] involved preparing transparent paper by coating fine paper with Venetian varnish and tracing the map. A form of carbon paper was then used to transfer the design to the plate. If the original drawing was expendable it could be varnished directly so that the drawing would show through and could be traced from the back.

When the transfer to the plate was completed, the paper was carefully removed in order to preserve the original drawing that was often the only copy available and would be needed by the engraver to refer to as he worked.

The method of transfer reported by Faithorne was obviously that used by Samuel Neele in 1786 when he wrote to Thomas Jefferson that "The drawing was unfortunately made on a paper much too soft, so that (after it was rubbed down on the wax prepared on the plate to receive its Impression previous to tracing) in taking it off, some parts of its surface tore away, & remained on the wax, & of course obliterated the drawing in those parts."[10]

In the late nineteenth century the Ordnance Survey spent considerable effort on experiments attempting to discover ways of plotting a map directly on to the copperplate from which it was to be printed, but this proved to be impractical.

Cutting. With the design transferred to the wax surface of the plate, the next step was that of incising the lines to be printed. This could be done either by etching or by engraving. In etching, a needle is used to scratch through a wax coating and acid is applied that eats away the metal to produce the lines to be printed. This process was not satisfactory for maps since an etched plate could not produce as many impressions as an engraved plate and the workman had less precise control over the lines to be printed. An example is found in the work of Wolfgang Lazius (fig. 3.2). For the most part, etching was reserved for decorative features. Some mapmakers used etching for marginal decorations such as is found on the maps in the Klencke Atlas. At other times it was used for the cartouche as illustrated by John Seller's *A Chart of the North Sea* prepared by James Clarke.

Line engraving produced the most satisfactory print. This was accomplished by running the point of a *graver* along the line to be printed to remove some of the metal and create a recess in the plate to hold the ink for printing. A skilled engraver could exercise a masterly control over his tool to produce a variety of effects. As Woodward notes:

There was considerable versatility in the thickness of line controlled by the engraver. The burin could be leaned slightly to one side to produce a thicker line, and such a leaning was natural in the engraving of curves, a slight thickening of line is often observable on curved engraved lines.

It is both in the thickening of lines at curves, and in the pointed character of the ends of lines that the copper line engraving can be distinguished from the etching, which has lines of relatively regular thickness and ends which are rounded.[11]

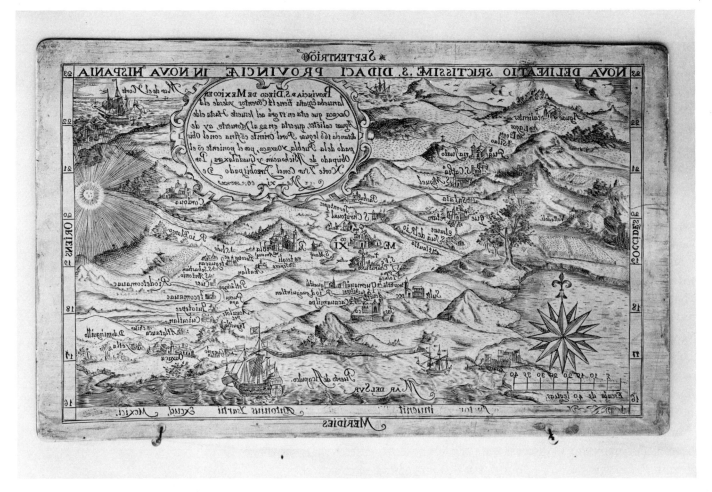

3.1 Original copperplate of a map of the province of San Diego, Mexico, by Antonio Ysarti, 1682.
Courtesy of Mr. and Mrs. Kenneth Nebenzahl

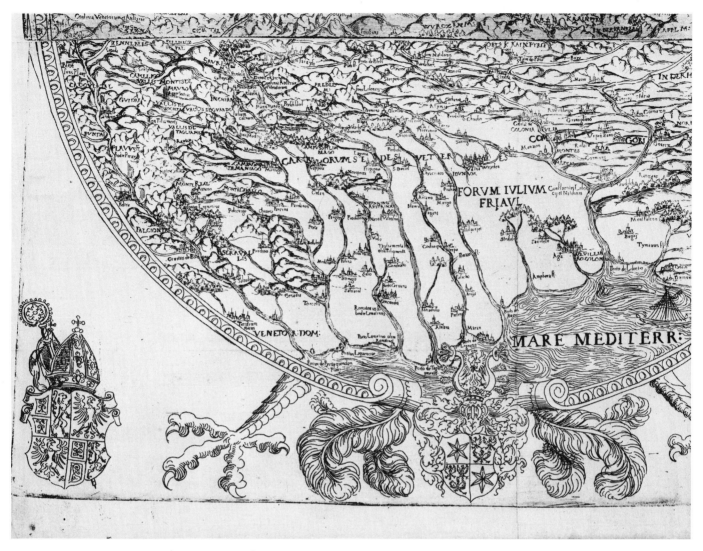

3.2 Wolfgang Lazius, *Carinthiae dvcatvs cvm palatinv Goricia* [Austria], ca. 1561.
Courtesy of the Newberry Library

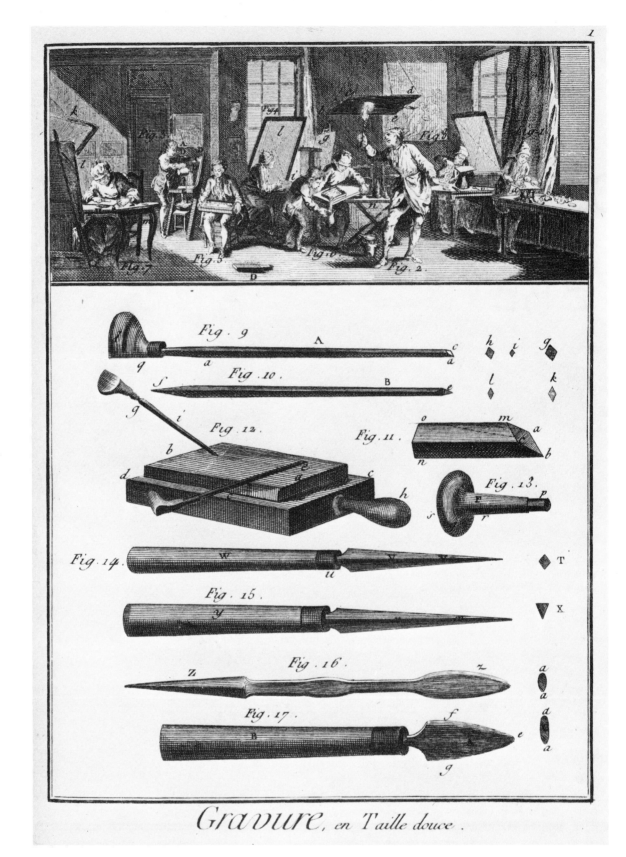

3.3 Engraving tools, from Denis Diderot, *Encyclopédie, ou dictionnaire raissonné des sciences, des arts et des métiers,* Paris, Briasson, 1751–65

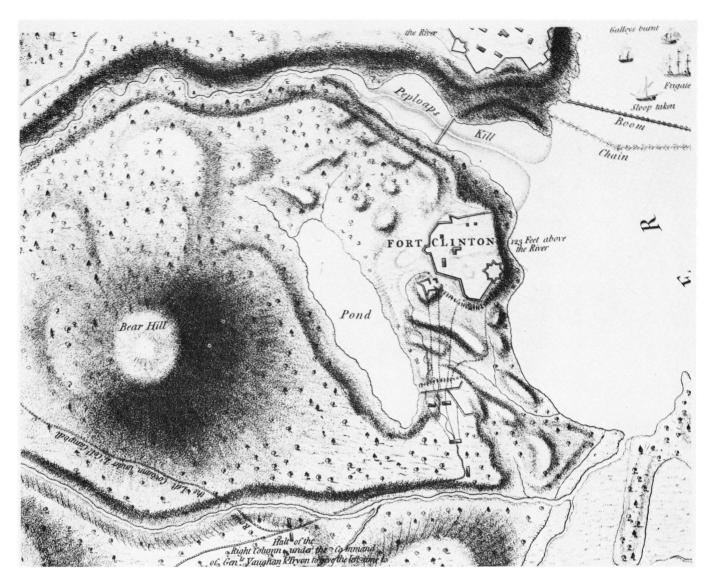

Labels within map: the River · Peploaps · Kill · Boom · Chain · Galleys burnt · Frigate · Sloop taken · FORT CLINTON · 123 Feet above the River · Pond · Bear Hill · the Left Column under Lt Coll Campbell · Halt of the Right Column under the Command of Genl Vaughan &c. given to have the left turn to · R.

3.4 Detail from John Hills, *Plan of the Attack of the Forts Clinton & Montgomery,*
London, William Faden, 1784.
Courtesy of the Newberry Library

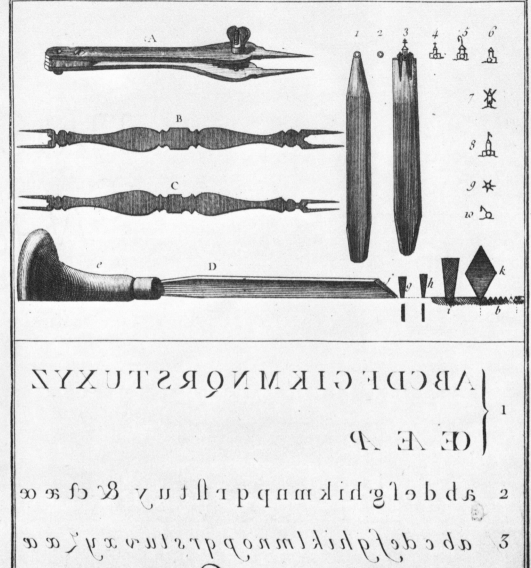

3.5 Punches and tools for engraving lettering.
From Denis Diderot, *Encyclopédie...*

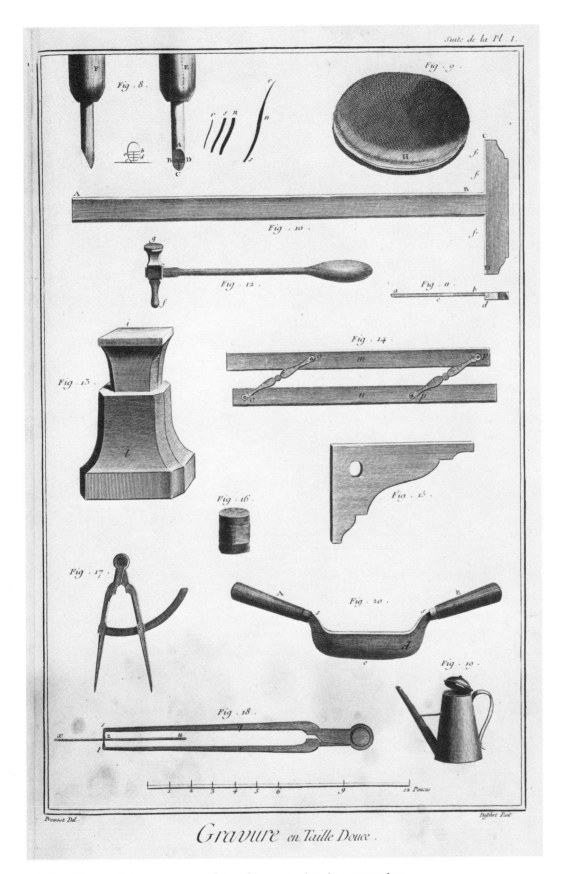

Gravure en *Taille Douce*.

3.6 Burnisher, anvil, hammer, etc., used in making corrections in a copperplate.
From Denis Diderot, *Encyclopédie*...

3.7a–b John Smith, *Virginia,* Oxford, 1612.
State 1. With detail showing ghost print.
Courtesy of the Ayer Collection, the Newberry Library

Ordinary howses 2 ────○

The Sasque=ahanougs
are a Gyant like peo=ple &
thus a=tyred

Cepowig

Vtchowig

Attaock

Tesinigh

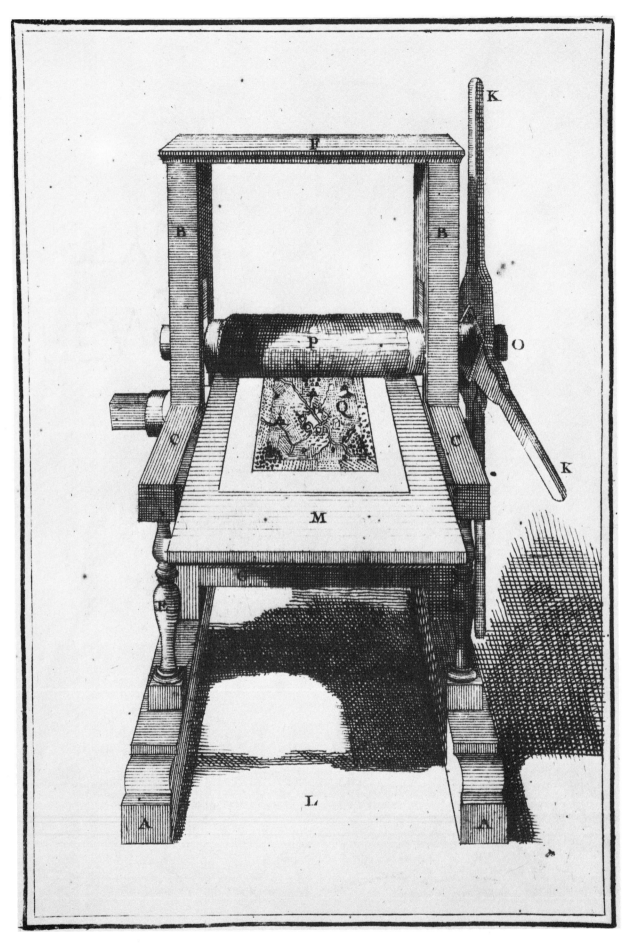

3.8 Rolling press, in Abraham Bosse, *De la manière de graver …,*
Paris, C. A. Jombert, 1745, plate 17

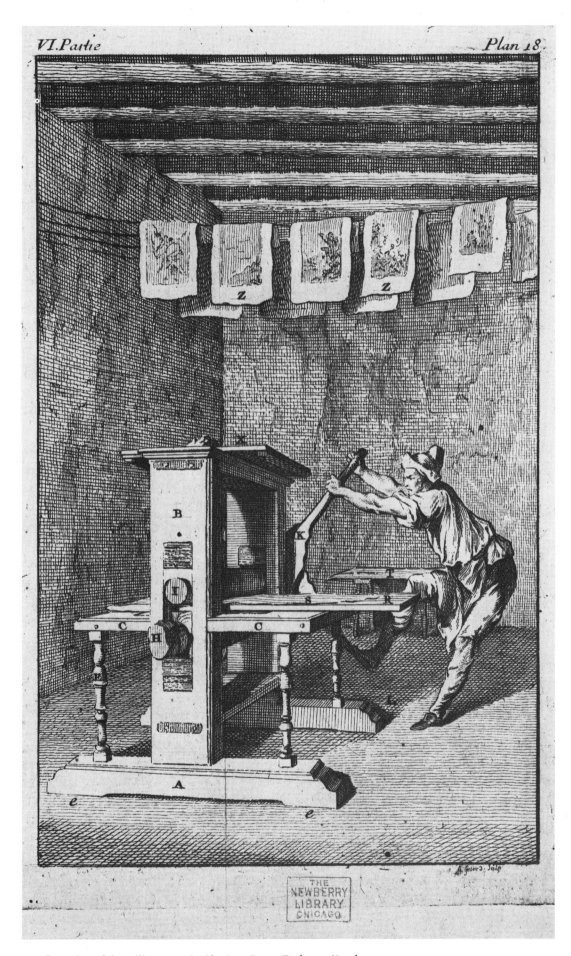

3.9 Operation of the rolling press, in Abraham Bosse, *De la manière de graver . . .,*
Paris, C. A. Jombert, 1745, plate 18

Sequence. In cutting the plate it was usual to follow a traditional sequence in the order in which certain elements in the design were cut. This was described by the Ordnance Survey: "The outline is first engraved, and then the writing, and afterwards the ornament consisting of rocks, woods, water, contours, &c. each class of work being usually done by a different person."[12]

This sequence in cutting can be observed on numerous extant maps. The map of Virginia published in the 1676 edition of Speed's *The Theatre of the Empire of Great Britain* has the delineation of the land area after John Smith but the toponymy after Augustine Herrman.[13] Various charts from Greenville Collins's *Coasting Pilot* (1693) show that the illustrative material was added after the lettering.[14]

While the plate was being cut, Faithorne recommended:

To the end you may better see that which is graven, they commonly role up close a piece of black Felt or Castor, liquored over a little with oyl-olive, and therewith rub the places graven: And if you perceive any scratches in your plate, rub them out with your burnisher: and if you have graved any of your strokes too deep, you make them appear fainter with rubbing them with your burnisher.[15]

In larger shops where several engravers were employed the different elements of the design were executed by different individuals. This may, in part, account for the many maps that do not include an engraver's signature. When Greenville Collins was preparing the charts from his survey of the coast of England, he engaged a number of different engravers, with some of the plates bearing one name for the geographical material and another for the decorative cartouche. The Robert de Vaugondy copy of the Fry and Jefferson map of Virginia is found signed by one engraver (E. Haussard) in the first state while another (Groux) signed *State 3.* Although the great Fry and Jefferson map is not signed, there is associated documentary evidence that indicates that the geographical material was cut by one or more hands while the cartouche was engraved by yet another engraver, named Grignon.[16]

Tools. The basic tool of the engraver is the *burin* or *graver.* This consists of a steel shank shaped and sharpened on one end with the other capped by a round woodblock. This block was shaped to fit snugly into the palm of the hand and the fingers stretched along the shank to control the cutting point. The tools in use today are not noticeably different from those in use in earlier periods (fig. 3.3).

There were two basic gravers that are described by Faithorne as follows:

. . . the one formed square, the other lozeng: the square Graver makes a broad and shallow stroak, or hatch; and the lozeng makes a deep and narrower stroke. The use of the square Graver is to make the largest strokes, and the use of the other is to make the strokes more delecite and lively.[17]

Some workmen designed and made gravers to their own specifications and, in time, a variety of specialized tools were created to produce special effects. Among such were the *tint tool,* the *scauper,* and the *threading tool.* These produced tone through shading and stippling. Occasionally, as in *The Atlantic Neptune* of J. F. W. Des Barres or in the battle plans of the American Revolutionary War by William Faden, a stippling effect was achieved with the roulette, a small wheel with spikes on its circumference, which was rolled back and forth over the copperplate (fig. 3.4).

Since there were conventional symbols to designate such things as towns or villages, some of these were made in the form of punches to produce symbols repeated frequently. The lettering on maps was usually cut by the graver, because of the variations in size and the space available (fig. 3.5).

The cutting portion of the graver had to be sharpened frequently and, as Faithorne advised, "It is very necessary that you take great care in the exact whetting of your Graver: for it is impossible that you should ever work with that neatnesse and curiosity as you desire, if your Graver be not very good, and rightly whetted."[18]

Time. The amount of time required to cut a plate depended upon the size and complexity of the map and the efficiency of the publishing establishment. In Augsburg, it took Kauffer three years to engrave the twenty-five plates for Muller's Bohemia published in 1722.[19] Samuel Hill of Boston completed the *Plan of the City of Washington* in something under two months. There were two virtually identical plates engraved, normally identified as the Philadelphia and Boston plates. Although the engravers of the Philadelphia plate said that they could engrave the plate in eight weeks, they actually took nine months, but this delay was due more to intransigence than anything else. Gregory King took three months in 1679 to engrave a two-sheet reduction of John Adams's large map of England and Wales (*Angliae totius tabulae* . . . 1677). The total size of the engraved surface of this map was 68.8 x 97.5 cm (27⅛ x 38¾ in).[20]

In the Preface to *Johnston's National Atlas,* published in 1844 and containing forty-six folio maps, is the note:

The intimate acquaintance of the Editor with the best methods of engraving, has secured for his drawings more than usuall exactness in their transfer to the plates, which have been entirely engraved within the last five years,—a much shorter period of time, it is believed, than has sufficed for the production of any Geographical work of equal extent . . . [and] . . . the employment of a succession of workmen almost day and night . . .

Cost. The cost of producing an engraved plate was considerable, and it varied with both the size of the plate and the complexity of the map. Gregory King charged John Adams £26.8s to engrave two plates.[21] Petty's *Ireland* cost £1,000 to be engraved in 1675 in Amsterdam.[22] In 1742, Emanuel Bowen indicated that a plate 69.0 x 56.0 cm (27¼ x 22⅛ in) would cost £10.10s for ". . . close work, such as the map of England . . ." and £3.10s for ". . . open work such as sea charts of ye coast."[23]

The price charged varied with the workman. William Faden quoted a price of 50 guineas (£52.10s) for Thomas Jefferson's plate (60.2 cm [24¾ in] square), but Samuel Neele actually cut it for £28.16s.9d.[24] In Philadelphia, in 1792, James Thackara and John Vallance charged $420 for a map of Washington, while Samuel Hill, in Boston, charged only $150 for an identical map only slightly smaller.[25]

In discussing the production of his great four-sheet map of Scotland, Aaron Arrowsmith noted: ". . . before I was in possession of the first impression (Five Hundred Copies) I had expended in Copper, Engraving, Paper, Printing and Colouring, £2,050, including about £100 lost by a cancelled plate . . ."[26]

His nephew John estimated that a plate prepared for the Royal Geographical Society for use in the *Journal* would cost about £5.15s for "drawing, cop[r], & engraving."[27] The thirteen maps by John Arrowsmith in volume 8 of the *Journal* cost £197. 3. 0. at an average cost of some £15. 3s.[28]

Proofs. At intervals during the cutting of the plate a proof is pulled in order to check the accuracy of the work in progress. The Ordnance Survey describes the procedure followed:

A proof is taken from the copper plate as soon as the outline and writing have been engraved, and after being carefully compared . . . is returned to the engraver with a list of errors and omissions. As soon as these have been attended to, the contours (etc.) are engraved, and a second proof is taken and examined, to see that the examiners remarks have been properly attended to, and to note any corrections required in the new work. The copper plate is then again corrected, the . . . [decorative material is engraved] . . . and everything is made complete, after which a third proof is taken and examined. Lastly, a fourth proof is printed and examined before the sheet is pronounced to be ready for publication.[29]

In order to facilitate the work of the engraver in making corrections, Faithorne suggests the use of a *counterproof,* which he describes as follows:

The Counterproof is made with the said Proof thus, *viz.* That having drawn the Proof put it smooth and all wet the wrong side on the plate which made it; then they put upon that Proof a Sheet of Paper wetted, then again the Bluring Paper over it, and lastly the Cloaths. They then pass all these together between the Rollers, and having taken up the said Sheet, they find that the Proof has made a Counterproof on the said Sheet of Paper. This is usually done to see better how to correct, because the said Counterproof is according to the Draught, *viz.* turn'd the same side.[30]

Although some proof impressions are known, no complete set of proofs for any one map is known to exist. If such were available it might be possible to analyze the cutting and correction of the plate in greater detail.

Corrections. The correction of errors in the incised material was relaitvely simple. Scratches or shallow lines were removed by rubbing the area with a burnisher. This spread the copper into the incised lines to be deleted so that they would not accept ink. Deep-cut lines or massive changes to one area on a plate were accomplished by pounding the copper from the back of the plate following the procedure described earlier in preparing the original plate. It would appear that the use of a burnisher on the surface was the easiest and most frequently used procedure for correcting errors (fig. 3.6). Both of these processes were also used when older plates were to be revised.

Neither the use of the burnisher nor beating always insured a permanent correction. If the plate had been printed so that ink remained in the lines this prevented a complete closing of the line. As the plate was worn with use, the original line was often worn open sufficiently to print faintly thereby producing a *ghost print* (fig. 3.7a and b). This situation is found in late impressions of the great Fry and Jefferson map of Virginia where the two western plates were deleted and new material engraved.[31] In several cases, Mount and Page used old copperplates for new maps without cleaning the plates with sufficient care to insure that no ghost printing occurred.[32]

Although there is little precise evidence of the full extent of proof corrections required in preparing a plate this must have been a major problem. Upon examining the proof of his map, Thomas Jefferson noted: "I have got thro about two thirds of the map and have a list of 172 errors, so that we may expect on the whole about

250."[33] The corrections were made by Guillaume Delahaye in Paris rather than by Samuel Neele, the original engraver. All in all, Jefferson had nineteen proofs pulled before he was satisfied and this did not include ". . . small and immaterial changes of orthography . . .," which he did not bother to correct. The printer charged ". . . 9 sols each paper and print . . ." for the proofs.

The frequency and care with which alterations were made to a copperplate influenced the length of the useful printing life of the plate.

The introduction of printing from incised metal plates produced a major alteration in the printing trade. When maps were printed from woodblocks the block was locked into the form with the standing type so that both text and illustrations were printed in a single operation at the press. Metal plates, on the other hand, required a different kind of press and a different procedure to produce the printed map. Although the similarities were numerous, the differences were sufficient to lead to the development of specialization in the production of prints from metal.

PRINTING

The press. Letterpress was printed on a flatbed press with static perpendicular pressure applied to a sheet of paper laid over the form of type. Metal plates, on the other hand, were printed on a roller press in which dynamic pressure was applied to the plate and paper progressively so that the ink was squeezed from the incised lines on the plate and transferred to the paper (fig. 3.8). Hansard describes the roller press as follows:

Copper-plate printing is done by a machine called the rolling-press, which may be considered as consisting of two parts; namely, the body and the carriage: analogous, in some respects, to the letter-press. The body consists of two cheeks of which the dimensions are not always the same; but, ordinarily, about four feet and a half high; a foot thick; and including a space of two and a half feet between them. These cheeks are fixed by tenons into wooden feet, which rest horizontally upon the floor, and to which they are made to stand perpendicular, being joined at the top by a cross-piece, or head. From the feet towards each end, rise also four other perpendicular pieces each about three feet high, which, with other horizontal pieces, form a frame-work both before and behind the cheeks, serving to sustain a smooth even plank about four feet and a half long, two feet and a half broad, and an inch and a half thick, upon which the engraven plate is to be placed for printing, and which may altogether be considered as the carriage of the press. Between the cheeks work two wooden cylinders or rollers, commonly of about six inches diameter, each of which has its ends lessened to about two inches diameter, called trunnions, by which each is borne in two pieces of wood in form of half-moons, and lined with polished iron to facilitate the motion, and which are made to play in grooves formed lengthways in the cheeks. The space in the half-moons left vacant by the trunnion is filled with paper, pasteboard, &c. so that they may be raised and lowered at discretion, to leave the necessary space between them for the passage of the plank charged with the plate, paper, and blankets. Lastly, to one of the trunnions of the upper roller is fastened a cross, consisting of two levers or bars of wood traversing each other. The arms of this cross serve in lieu of the handle of the common press, by the aid of which the workman gives a motion to the upper roller, by which a motion being also communicated to the under one, the plank is forced backwards and forwards between them, with the plate and paper, and hence impressions are obtained.[34]

The ink. The preparation of ink for printing involved grinding the pigment and its suspension in an oil base of suitable consistency. The best pigment was German Black produced in Frankfurt. Both Faithorne and the Ordnance Survey recommended this particular pigment as the most desirable. Faithorne noted:

> Its Goodness and Beauty consists in having an Eye and Black of Velvet, and that rubbing it in your Finguers it squeezeth like fine Chalk or raw Starch. The Counterfeit has not such a beautiful Black, and instead of feeling soft to the Finguers, 'tis ruff and sandy, and weareth out the Plates, 'Tis made of the dregs of burnt Wine.[35]

The finely ground pigment was suspended in a medium of nut oil, which was carefully prepared. The oil was placed in an iron pot and heated until it boiled when it was then ignited and allowed to burn off. Burning for at least half an hour produced a *weak oil*. A second pot of oil was burned longer until it became ". . . very thick and clammy . . ." to make *strong oil*. Both kinds of oil were required in mixing the ink.

The prepared pigment was combined with some *weak oil* very slowly and ground as dry as possible. This was then mixed with some *strong oil* and ground again until it was ". . . extreme roping and clammy, like a very thick Syrup."

Newly engraved plates with deep incisions required a thicker ink made of more of the strong oil but worn plates or those not cut deeply required less of the strong oil.

The paper. The paper to be used in printing needed to be dampened to insure that it would melt into the incised lines to absorb the ink from the plate. Each sheet of paper was passed through clean water two or three times depending on the thickness of the paper. It was then laid on a smooth board. When all of the required sheets had been dampened and stacked together a second smooth board was placed atop the pile and heavily weighted with stones. This ensured an even dampening of all the sheets and forced out any excess water. By the next morning the paper was ready for use. "Strong and well gum'd Paper must be more wet, and the weaker and less gum'd less," according to Faithorne.

Some kinds of paper would shrink more than others so that different impressions from a single plate may be found to vary slightly in size. Although this variation may not be pronounced, it could be sufficient to mislead the unwary into assuming that different plates of an identical map were used.

Inking. In preparation for inking, an inking ball is made ". . . of good soft and fine Linnen half worn out . . ."[36] rolled into a close hard ball about five inches in diameter and three inches thick, and the engraved plate is heated "pretty warm." The ball is dipped in the ink and by ". . . sliding, pressing, dappling in sundry ways . . . over all the Engraved Surface of the Plate, you make the Black enter and pierce into all . . ."[37] the incised lines. When this is completed the surface of the plate is carefully wiped clean with a soft rag. A further cleaning is achieved by rubbing the plate with a clean rag on the ball of the hand until the plate is polished except for the ink trapped in the incised lines.

A charcoal brazier was used to heat the plate while it was being inked. Hansard noted that "This is a part of the process which renders the business extremely injurious to the health of the workmen, in consequence of the noxious vapour arising from the charcoal. . . ."[38] In 1818, James Ramshaw invented a steam box to replace the brazier and was awarded a gold medal by the Society for the Encouragement of Arts.

Presswork. After inking and cleaning the plate is again heated until it is warm:

The plate, thus prepared, is next laid on the plank of the press, and upon it is placed the paper, well moistened after the manner described. . . . Two or three folds of flannel are then brought over the plate, and things thus disposed, the press is set in motion by pulling the arms of the cross, by which means the plank bearing the plate and paper is carried through between the rollers, which, pinching very forcibly and equally, press the moistened and yielding paper into the strokes of the engraver, whence it draws out a sufficient portion of the ink to display every line of the intended print.[39]

The passage of the plate between the rollers must be made ". . . gently and roundly, and not by jolts and joggs, that the Print clean without Blurs, Spots, or Wrinkles . . ." (fig. 3.9).[40] When the plate has passed through the rollers, the paper is removed and hung to dry and the plate is inked again for the next print.

A large shop would have crews working at each side of the press inking plates and two plates were usually worked at the press simultaneously. A normal press run consisted of one hundred impressions in a day. In 1742, Emanuel Bowen noted that one hundred was a day's work at a cost of six shillings.[41] In 1792, in Philadelphia, Mr. Scott indicated that the same number of impressions in a day cost $3.33 while John Arrowsmith in 1847 noted his cost at six shillings for a hundred in a day.[42]

Drying and cleaning. The printed paper is hung to dry with weights attached at the bottom to prevent the paper from wrinkling as it drys. After the day's run is finished the prints are stacked with interleaves to protect the printed surface, which dries more slowly than the paper, and the stack is weighted with rocks.

The plate is cleaned carefully with olive oil to remove the ink from the incised lines and then wrapped in clean paper for storage. .

Color. The addition of color to a printed map was always considered an asset as much for its decorative quality as for its utility in transmitting information. Color was generally added to an impression by hand. The art of coloring maps developed as a subspecialty in the trade either within a map publishing house or independently. A map of Canada published in 1643 has an imprint reading ". . . chez Jean Boisseau, Enlumineur du Roy . . ." indicating that this aspect of the industry had achieved some recognition. In 1787, Jefferson paid the bookseller Froullé in Paris "avoir fait colorier 440 cartes de la Virginie 7' 10 . . . 33."[43]

The application of color to an impression by hand was a fairly common practice, and certain conventions were established with respect to the use of color. The best known work on the art of coloring maps is contained in a book by John Smith published in 1769 that not only describes the conventional code but also provides instructions for mixing and applying color to printed maps.[44] As the geographical content of maps became more precise, the use of color became increasingly more restricted.

Attempts were made to print in color from engraved plates. Faithorne described the method then in use, which involved the preparation of one or more identical plates with each having incised on it only those parts to be printed in a single color, so that after the careful registry and superimposition of each plate a color print was produced.[45] No maps printed in color from engraved metal plates are known until the eighteenth century. Woodward identifies one example from Jefferys's *General Topography* published in 1768,[46] and Bellin's *Carte Réduite des Isles Philippines* (1752) has a dark yellow rhumb line grid printed from a separate copperplate.

In 1845, the *Journal* of the Royal Geographical Society noted:

Colour has long been considered a powerful auxiliary to mere engraving in maps of every kind; indeed, for certain special purposes, as in geological maps, it is indispensable. Any other mode of colouring such maps than by hand, has always been considered very difficult, for some reasons sufficiently evident of themselves, and others which it would be too long for me to explain in this place. It is therefore with great pleasure we learn that M. Dufrenoy has presented to the Academy of Sciences a "Memoir on the Colouring of Maps," by which all the defects hitherto inherent on the application of tints, by mechanical means, have been corrected; and the greatest complication of colouring, as in the case of the great Geological Map of France, by himself and M. Elie de Beaumont, is now effected with perfect ease and accuracy, and at a very considerable reduction of cost.[47]

The coloring of maps by mechanical means was not successfully achieved until the advent of surface printing processes such as chromolithography in the nineteenth century.

THE INDUSTRY   The desirability of copperplate engraving for printing maps resulted in the development of a new industry that had highly specialized roles for its members. This aspect of the history of mapmaking has been largely neglected by historians, consequently there is little precise evidence to describe the structural patterns of the industry or the work roles of its personnel. Robinson has discussed in chapter 1 the relationships that developed between the cartographer and the publisher so this discussion is concerned with some aspects of the publishing industry.

Structure. As shown above, the production of a map printed from an engraved metal plate involved a number of different tasks. An individual entrepreneur could and did perform every task required to produce a map, but specialization of roles occurred so that a number of constituent specialties developed that often became independent establishments.

From the scant data available it appears that there were several distinct patterns for the publication of maps that can be catalogued in terms of the primary role in the trade.

Cartography. Every map begins with the cartographer who can present geographical information in visual form. The cartographer usually prepared the original draft, which was then copied for the printer by draftsmen. Some cartographers are known only by the manuscript maps and charts that they made, and many of these never appeared in printed form. It is only recently that research has begun to clarify the role of the cartographer in publishing maps. One famous group of cartographers, including John Daniel, Nicholas Comberford, and John Thornton, existed under the aegis of the Drapers' Company.[48]

Some noted cartographers were also engravers and published their own charts. This pattern may be considered to be the origin of the map and chart trade and is found to persist in every century. Men like Gerhard Mercator, John Thornton, Herman Moll, and John Arrowsmith were equally skilled as cartographers and engravers, so that they could publish their own work.

In many cases, shops established by cartographers were self-sufficient with respect to the production of maps. John Rocque is known to have had ten men in his shop including ". . . as many draughtsmen as engravers. . . ." In spite of having their own craftsmen at hand, some cartographers found it necessary to engage independent engravers to cut plates. In discussing his map of Scotland published in 1807, Aaron Arrowsmith noted: "I was also careful to employ one and the same person in etching the whole Map . . . and it is right to say, in justice to Mr. John Smith of Chelsea, that he performed this to my satisfaction."[49]

It was not often that the cartographer could also handle letterpress printing. Where this was essential it was usually done by an established printer. John Seller employed John Darby to print the text for his first marine atlases, and John Thornton used John How for his *Oriental Navigation* published in 1703. In some cases the cartographer and printer combined their resources to produce atlases as did John Thornton and William Fisher when they published the *Fourth Book of The English Pilot*.[50] Eventually, some firms started by cartographers grew to where they could handle all aspects of printing and publishing as is characteristic of modern firms such as Bartholomews or Stanfords.

Engraving. Although some firms had their own engravers, there were also many engravers who maintained independent establishments and worked for mapmakers as well as others. Wenceslaus Hollar made some map plates such as those he cut for John Seller, but he is best known for his prints. Faithorne engraved prints primarily, but he also cut the four plates for the Hermann map of Virginia. Samuel Neele operated his own establishment in London and advertised that he was available for "Engraving in general . . . performed with neatness and expedition." Neele engraved map plates for John Stockdale, among others.

It was not unusual for an engraver to begin working in a cartographer's shop and then move out on his own. In some cases they became important publishers of cartographic material. George Frederick Cruchley advertised himself as "From Arrowsmiths" and eventually took over the John Cary map firm.[51]

Printing. Although the map printing trade began with cartographers who were also engravers and publishers and this pattern persisted through the centuries, there were some publishers with no skills in the production of printed maps. These firms usually started with letterpress printing and moved into maps and charts in order to consolidate both letterpress text and map printing for geography books, atlases, sailing directions, or marine atlases. In general, these firms depended upon independent cartographers for the drafts of the maps they published but maintained their own staff of engravers and printers. One of the longest tenured firms of this type was that of Mount and Page, which began as such about 1700 and continued in business through the whole of the eighteenth century.[52] This firm published books, atlases, and separate charts related to navigation and acquired their stock of map plates largely by purchase from other publishers although they did produce some plates of their own. They maintained a complete establishment capable of producing everything from letterpress to engraved plates including the final binding of the finished product. Mount and Page were the forerunners of modern establishments like Rand McNally.

Practices. The engraved copperplate had a flexibility that led the map trade into many curious practices that are not yet identified or clearly understood. One major concern of the publisher was that of prolonging the useful life of the plate.

The useful life of a copperplate cannot be asserted with conviction since there is virtually no reliable data to indicate the number of impressions that could be pulled from a given plate. According to Skelton: "It is estimated that 2,000 to 3,000 impressions might be taken from a plate or block without serious wear; yet by careful husbandry and (if necessary) by retouching of the incised lines the life of a copper plate could be and often was, prolonged to a phenomenal span."[53] On the other hand, Vittorio da Zonca[54] noted in 1607 that a copperplate would provide one thousand impressions when carefully used and the Ordnance Survey indicated that "A copperplate with fine work upon it will generally give about 500 good impressions."[55]

There are many maps known to have been printed at intervals over a phenomenal span of time. Some Speed maps were printed from the same plates from 1611 to 1770, yet this is not necessarily reliable evidence since the number of impressions pulled from the plates is not known.[56] A plate lightly used could well provide usable impressions for well over a century while one used extensively might last but a short time. The plate made for Thomas Jefferson's *Notes on Virginia* is known to have produced some 2,925 impressions in two years without evidence of recutting observed in later copies, but plate wear is obvious.

The procedures followed in preparing the plate originally and in the printing would materially influence plate life so that only the very best plates correctly handled could be expected to have a lengthy usable life. An uneven surface resulting from inadequate polishing would limit the number of successful impressions from a plate. Too many corrections or alterations would weaken the copper so that it could not survive extensive use. One of the plates in Collins's *Coasting Pilot* was damaged in the bottom left quarter so that a piece 19.8 x 24.8 cm (7.8 x 9.8 in.) was removed and a new piece substituted. From 1693 to 1753 this chart was printed from these two plates, which undoubtedly complicated the presswork.

Perhaps the most important matter was that of presswork. Improper pressure on the rollers, an uneven pull, or inadequate padding over the plate would result in plate damage. Mark Tiddeman's chart of New York Harbor was mishandled in the press so that the plate was cracked, and this crack increased to the point where the plate was no longer usable.[57]

When plate wear reached the stage where legible impressions were no longer printed, it was possible to prolong plate life somewhat by recutting the incised lines. In most cases this is readily observed in an impression, but it is within the realm of possibility that a very skilled workman could recut a plate and it not be detectable. Plates tended to wear unevenly because of variations in the depth of the incised material; consequently, only parts of the plate may need recutting at any one time. No map is known in which the entire plate was recut at one and the same time. Fading occurred first in those areas that were stippled or where very fine lines were used as in shaded areas or small lettering.

Some printers were very attentive to the quality of the impressions and recut their plates often. Others were indifferent. Mount and Page were notoriously unconcerned about the quality of the impressions of the charts in their *English Pilot* series, consequently some plates were used so often that the primary matter was scarcely more legible than an accidental ghost print. Stippled shoals and sand banks were allowed to fade completely so that when the plate was recut the cartobibliographer could be misled into thinking that new geographical data had been added.[58]

There was a limit, of course, to the amount of recutting that was possible in an effort to prolong plate life. The constant heating of the plate during inking and the tremendous pressure exerted by the rollers would weaken the metal to such an extent that a plate could not be revived. In such a situation it would have to be abandoned or a new plate made. In most cases a wholly new map would replace the old, but some publishers such as Mount and Page merely cut a new plate of the old map.

A different approach to prolonging plate life was introduced in the nineteenth century with the use of *electrotyping*. In 1842, the president of the Royal Geographical Society noted that

The art of multiplying maps by means of electrotype plates is making considerable progress in Germany. In Saxony it has been attended with great success, and its practical application is likely to lead to the extension of geographical, or at least topographical knowledge: . . . it is quite impossible, when done with care, for the most experienced eye to perceive the slightest difference between an impression from the original engraved plate and one from the electrotype plate.[59]

The Ordnance Survey also used electrotyping to produce facsimiles of engraved copperplates during the latter part of the nineteenth century. The question of the detection of impressions from an original plate and those from a facsimile plate has not been studied. In 1857, the process of steel-facing of a copperplate was discovered that reduced wear and prolonged plate life almost indefinitely. This is in use today principally in plates made for popular prints that are issued as restrikes from the original plate. No early maps are known to be available as restrikes.

Publishers altered plates in other ways for reasons not associated with prolonging plate life. It was not uncommon for a publisher to cut up a copperplate and use the parts separately. Mount and Page divided Thornton's original folio plate containing maps of Bermuda and Barbados so that each could be printed separately in-text. John Seller trimmed the old Dutch plates he bought for his *English Pilot* so that they would fit conveniently into his folio atlas. Jansson, Hondius, Visscher, and Blaeu are all known to have removed the original decorative borders from plates so that they would fit into an atlas.

Along with trimming plates, publishers also added plates. Thomas Jefferys added a supplemental plate measuring 27.7 cm (10¾ in.) to the left side of his map of *Canada and the Northpart of Louisiana* to include a western extension of the geographical data. Aaron Arrowsmith trimmed each side of his great map of the *Interior parts of North America* and added two additional plates on the bottom so that he could include the whole of North America rather than just the northern parts. Additions were also made in the form of small slips printed separately and designed to be pasted over a portion of an impression from a plate. Arrowsmith followed this procedure to add the Hearne data on copies in stock of the first state of his map of *Interior parts . . .,* and Des Barres did the same on his map of Yorktown.[60]

The map of Brunswick in Blaeu's rare *Appendix* of 1631 has the imprint and scale printed from different plates, while Delisle's map of South America has the title and cartouche printed from a separate plate cut to fit into a hole in the primary plate.[61]

At the time of printing it was not unusual for the publisher to mask portions of the plate so that false plate marks occur. This is observed in Doncker's map of Australia (*Wester Deel vom Oost Indien . . .*) published in 1660.

These and other curious practices not specified do indeed make the early printed map a bibliographical monstrosity and emphasizes the growing recognition of the importance of the study of form as well as content in the history of cartography.

RESEARCH The nature of copperplate printing and the publishing practices that developed around it have created the need for a systematic approach to the study of early maps as objects. In the past, as Robinson has indicated, scholars ". . . have been more concerned with the geographical information on the maps than they have with the technologies involved in their preparation. . . ."[62] Consequently, although there is an abundance of research detailing the evolution of geographical knowledge, there is very little known about the principal instruments through which that knowledge was diffused. Furthermore, such research as has been done about map publishing has been largely casual, somewhat indifferent in scholarship, and unsystematic. This has led to gross misconceptions about map publishing and the persistence of errors of fact through scholarly incest.

Sir Herbert George Fordham could be credited with introducing the study of the map as an object, but it is only in very recent years that the literature of the history of cartography has begun to include reliable studies of the publishing phase.

Because the study of map printing is of recent interest and because there is much to be learned about the application of copperplate printing in the map trade, there is a danger in wandering too far afield rather than concentrating on the immediately relevant aspects of the subject. The technologies of copper engraving discussed here— including matters pertaining to the metal used, the tools, and the engraving process itself—are not unique to maps and are, therefore, not directly relevant. The ways in which mapmakers adapted the process to their own ends is supremely relevant. Thus, the history of engraving can be left to specialists in that subject. A similar situation exists with respect to the history of paper including its production, distribution, and watermarks.

The principal areas of research needed to understand and clarify the history of copperplate engraving as applied to the production and distribution of maps would include the following:

Research methodology. Although Fordham inaugurated the systematic study of maps, there has been no outstanding progress since in the establishment of systematic criteria for the study and description of maps. As a result, the voluminous literature about maps is inconsistent with respect to the information reported, the terminology used, or the procedures employed in the analysis of the data. In most cases, a map described in the literature cannot be positively related to a copy in hand, which is essential in order to discover variations from the norm.[63]

Plate histories. One of the most useful pursuits in the study of maps is the careful, systematic, and detailed history of a single map plate that includes the original map and its direct derivations. Plate histories combine historical data with descriptive data in such a way that the origin of the map and its publishing history are carefully delineated. Such histories usually provide crucial data useful for secondary analysis and are the basic source material for other kinds of needed research. Included in this category would be studies of single atlases and the constituent plates.[64]

Style. Both Fordham and Lynam have demonstrated the interest and value present in questions of style and the evolution of style through the ages.[65] De Dainville's work on *Le Langage des Géographes*[66] is important on this subject, but it is not concerned primarily with style and does not provide an adequate systematic treatment of the evolution of style. This subject has particular utility to the matter of identifying and dating maps, and would be basic to cartobibliographical description.

Publishing. There is scanty evidence about the map publishing industry as discussed earlier. Research is needed to describe and analyze the structure of and the procedures that operated in publishing houses at various times in history. The Blaeu and Hondius establishments dominated the seventeenth century, as did Arrowsmith in the nineteenth. In what ways were they similar or different with respect to their internal operations? Studies of such firms would help answer interesting questions related to the time and cost of producing a plate as well as the question of plate life.

Marketing. The study of marketing practices such as the interrelationships among contemporary publishers would do much to clarify the role of map publishers and vendors in the diffusion of geographical knowledge. Although there is some literature related to this question, there are too few detailed studies to explain some puzzling anomalies encountered in the spread of information.

Publishing practices. Nearly every cartobibliographical study mentions some curious condition encountered in early printed maps yet there has been little systematic study of the "curious and forgotten lore" in the ways publishers used their plates. No doubt such research would produce some surprising results and dispel many cherished myths of historians of cartography.

Dating. There is scarcely any subject as crucial to the history of cartography as the date of a given map, consequently, there is a pressing need for systematic research into the numerous characteristics of maps and map printing that would be clues to the more precise assignment of tentative dates. At the moment, dates are often assigned on subjective grounds due largely to the lack of reliable research that would permit greater accuracy. Such matters as the invention and introduction of the various projections, the introduction of symbolic conventions, the sequence of various cartouche designs, or the matter of scales, could materially influence accuracy in dating and even correct some dates now in current use.

Biographical studies. Far too little is known about the individuals and firms that have been part of the map trade. Detailed biographies and histories of both, whether great or small, will be extremely useful. Fordham's study of John Cary and the biographies of Mercator or Blaeu are examples of work in this area.[67]

4 | Lithography and maps, 1796-1850

Walter W. Ristow

INTRODUCTION

The contribution of engraving to the history and development of cartography was outstanding. Nonetheless, there were certain limitations to engraving, which was a highly skilled, laborious, and costly technique. It is little wonder, therefore, that, until well into the nineteenth century, maps and atlases, because of their cost and scarcity, were available to only the more affluent members of society. During the latter half of the eighteenth century various individuals sought to devise more expeditious and less costly techniques for duplicating maps and other graphics. First to achieve practical results was Alois Senefelder who, in 1796, discovered lithography. The birth and early development of the new art are well documented. In *Vollständiges Lehrbuch der Steindruckerey . . .,* published in 1818, Senefelder detailed the events that led to his significant discovery and described the several operations involved in lithographic printing. Following publication, in 1819, of English and French translations of Senefelder's book, the new reproduction technique experienced rapid growth and development.

Senefelder conducted numerous experiments relating to materials, equipment, and applications of the process during the two decades that elapsed between the invention of lithography and publication of his book. By 1818 he had learned that metal plates or a grained paper could be substituted for stone, experimented with color printing, developed a special paper from which images could be transferred to stone, thereby making it unnecessary to draw or write in reverse, printed from polished, engraved, or slightly etched surfaces, and devised several types of printing presses. In his *Complete Course,* the inventor declared, with apparent justification, that

beside myself . . . no person, in all the different branches of Lithography, has effected any new improvement of consequence, which he had not received directly, or indirectly, from me; that all those artists, and producers of prints, made their first essays under my immediate direction, or were trained and instructed by persons who derived their information from my instructions.[1]

Lithography, in contrast to older printing methods, is a planographic process. The lithographic image is transferred from a smooth, or nearly smooth, surface to the paper rather than from a raised (woodcut or movable type) or incised surface (engraved plate). The transfer is effected by chemical, rather than physical, action. Senefelder, in fact, initially referred to his invention as chemical printing, and he considered lithography, or printing from stone plates, as but one application of the process. Nonetheless, the designation *lithography,* introduced in France as early as 1804 (French, *lithographie*), came to be generally accepted, and in 1819 Senefelder incorporated it in the title of the English version of his book.

The experiments that led to Senefelder's invention of lithography began about 1796 when he observed that slabs of Solenhofen limestone, quarried not far from Munich, attracted both water and grease. He carried on further experiments utilizing a greasy ink, prepared by mixing wax, soap, and lampblack in proper proportions, with which he applied images directly to the polished stone surface or transferred them from paper to stone. Explaining how lithography differed from printing procedures that relied upon elevated or engraved surfaces, Senefelder noted that

the chemical process of printing is totally different from both. Here [i.e., in lithography] it does not matter whether the lines be engraved or elevated; but the lines and points to be printed ought to be covered with a liquid, to which the ink, consisting of a homogeneous substance, must adhere, according to its chemical affinity and the laws of attraction, while, at the same time, all those places that are to remain blank, must possess the quality of repelling the colour. These two conditions, of a purely chemical nature, are perfectly attained by the chemical process of printing; for common experience shews that all greasy substances . . . or such as are easily soluble in oil . . . do not unite with any watery liquid, without the intervention of a connecting medium; but that, on the contrary, they are inimical to water, and seem to repel it. . . . Upon this experience [the inventor emphasized] rests the whole foundation of the new method of printing . . . because the reason why the ink, prepared of a sebaceous matter, adheres only to the lines drawn on the plate, and is repelled from the rest of the wetted surface, depends entirely on the mutual chemical affinity, and not on mechanical contact alone.[2]

Senefelder summarized the entire lithographic process as follows:

to wash the polished stone with soap-water, to dry it well, to write or draw upon it with the composition ink or soap and wax, then to etch it with aqua-fortis [i.e., nitric acid] and, lastly, to prepare it for printing with an infusion of gum-water. . . . The gum-water . . . [he explained] enters into affinity with the stone, and stops its pores still more effectually against the fat and opens them to the water.[3]

As with any new invention or process, lithography was subjected to much experimentation and exploration by many individuals during its early years. Notwithstanding the dissemination of information by the inventor and his disciples, prior to publication of Senefelder's *Complete Course of Lithography,* the corpus of knowledge relating to lithography was unorganized and unsystematic. There was, during this early period, little agreement as to what constituted the lithographic process. As Twyman observes, "what we now know as lithography began as a number of imitative processes whose only common factor was the lithographic stone. . . . Sometimes the stone was etched or incised to imitate a wood engraving, and sometimes it was etched or engraved as on copper. . . ."[4]

Perfection of the lithographic process was greatly dependent upon the development of effective printing presses. No one was more sensitive to this need than Senefelder. "I think myself bound to declare," he wrote,

that I deem it one of the most essential imperfections of Lithography, that the beauty and quantity of the impressions, in a great measure, depend on the skill and assiduity of the printer. . . . Till the voluntary action of the human hand is no longer necessary and till the impression can be produced wholly by good machinery, I shall not believe that the art of Lithography has approached its highest perfection.[5]

Special presses were constructed by Senefelder and his contemporaries, but, Twyman notes, "most of the earliest lithographers . . . seem to have begun by adapting the familiar presses of the letterpress and copperplate printer."[6] He acknowledges that "apart from Senefelder's pole press and a few portable models, we know very little about the earliest lithographic presses and next to nothing about their actual appearance."[7] The inventor claimed that he could take twelve hundred prints a day on his pole press, but Twyman believes that for large sheets four hundred impressions may have been a more realistic daily output.[8]

Despite the paucity of published information about lithography, there was considerable interest in the new process during the early decades of the nineteenth century. A number of individuals traveled to Munich to learn about lithography from Alois Senefelder, his brothers, or other early practitioners. Some returned to their places of residence and established lithographic printing plants. By 1820 the new process had been introduced, with varying degrees of success and continuity, in some fifteen or twenty European cities. While maps were among the materials reproduced in the early years of lithography, few lithographic plants were exclusively concerned with cartographic printing. For the first two decades following Senefelder's invention, therefore, the development of cartographic lithography coincides with the general history of the process.

Notwithstanding Senefelder's detailed account of his discovery, much in the early history of the technique is obscure. This is particularly true with respect to applications of lithography to map printing. We are not able, for example, to identify the first map reproduced from a stone, or to establish when this noteworthy event in cartographic history transpired. We may conjecture that the ancestor of all lithographic maps portrayed a limited area, probably of a city or a portion of it. Confirmation of this is found in Luitpold Düssler's *Die Incunabeln der deutschen Lithographie (1796–1821)*.[9] Of the forty or so titles that describe maps, more than 60 percent are of German cities. Several date from as early as 1808. Most maps printed by lithography during the incunable period were probably issued in small editions that had limited distribution. Only a few examples are preserved in archives and libraries.

In the course of his career, Senefelder was associated with various individuals in seeking to promote the lithographic process. Thus, late in 1806, he entered into a partnership with Baron Christoph Aretin that continued for about four years. During this period, as related in his *Complete Course,* Senefelder "executed a great variety of works, such as circulars, tables, and other things, for the government; plans, maps, and works of art, in the different manners."[10] However, he does not indicate when he produced his first lithographic map. Senefelder's greatest involvement in map printing began in 1809. The kingdom of Bavaria established, by royal decree in 1808, the

THE INCUNABLE PERIOD IN
CARTOGRAPHIC LITHOGRAPHY

Steuer Kataster Kommission (Cadastral Survey), which was charged with preparing detailed maps to regulate land apportionment, ownership, and taxation. Because the surveys were made at the scale of 1:5,000 (1:2,500 for urban areas), printing the numerous sheets was a major undertaking. Joseph von Utschneider, director of the project, wisely decided to reproduce the maps by lithography. The lithographic plant was organized under the direction of Johann M. Mettenleiter, who had learned the process from Senefelder some years before. During the period of its maximum operation some thirty or forty lithographic engravers were employed. In October 1809 Alois Senefelder was appointed superintendent of lithographic printing of the Cadastral Survey, a post he filled until his retirement in 1827. More than twenty thousand different map sheets were ultimately printed by the Steuer Kataster Kommission during the period 1809 to 1853.

The reproduction process employed by the Cadastral Survey was an adaptation of stone engraving. That is, the image was cut into the surface of the stone with a needle, for Senefelder believed that this method had advantages for map printing. Describing the "engraved manner of lithography" in his *Complete Course,* the inventor explained that

the greasy substance, which, on the stone, is to attract the printing ink, lies beneath its surface, as the lines of the writing, or drawing, are previously engraved into the stone by the needle, or etched into it with an acid, and subsequently filled with grease. . . . Like the elevated manner, it has several subdivisions. This manner [Senefelder wrote] is one of the most useful in Lithography, and is nearly equal to the best copper-plate printing . . . while the stone can be wrought rather more expeditiously and easily than copper; for fine writing and maps, it is peculiarly well adapted, as the number of lithographic maps published sufficiently prove.[11]

Twyman believes that "by 1825 the planographic method seems to have been generally accepted as the most natural method of printing from stone, except for specific types of work like lettering, mechanical drawing, maps, and architectural plans, where greater precision was required."[12]

Military historical cartography was another division of mapmaking that early utilized the lithographic process. In the Library of Congress Geography and Map Division there is an imperfect atlas, lacking a title page, with some sixteen plates of maps and plans that illustrate battles of the War of the Austrian Succession and the Seven Years' War. All the maps were reproduced by stone engraving at Munich in 1810 or 1811 (fig. 4.1). Troop positions are hand-colored to differentiate the opposing armies.

Munich, the birthplace of lithography, continued to be the most important center for the new process for the first decade or so after its invention. As early as 1809 there were seven lithographic printing houses in the city. There was, however, universal interest in the new, low-cost reproduction technique. Senefelder was generous in explaining his process to visitors from within Germany as well as from other countries, and only a few years after the invention of the process lithographic shops were operating in several European countries.

In 1801, Senefelder went to London, accompanied by Philippe André, brother of one of the inventor's partners. An English patent, applied for by André in Senefelder's name, was obtained for the lithographic process. Little progress was made, however, in exploiting the invention in England and, in less than a year, Senefelder re-

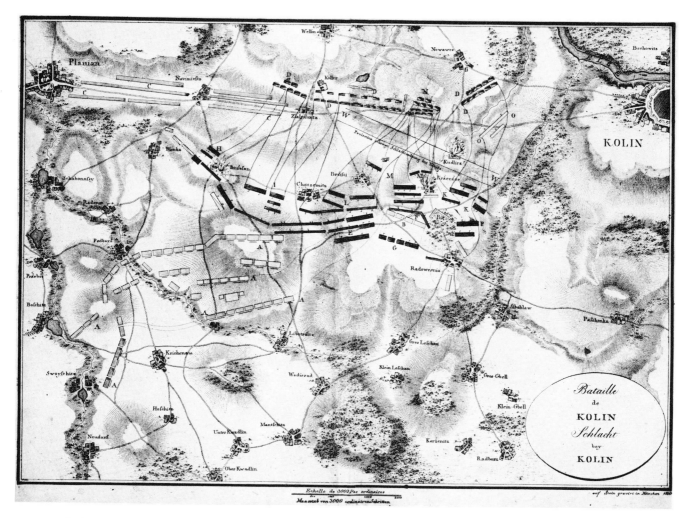

4.1 *Bataille de Kolin,* Munich, 1810.
Courtesy of the Library of Congress

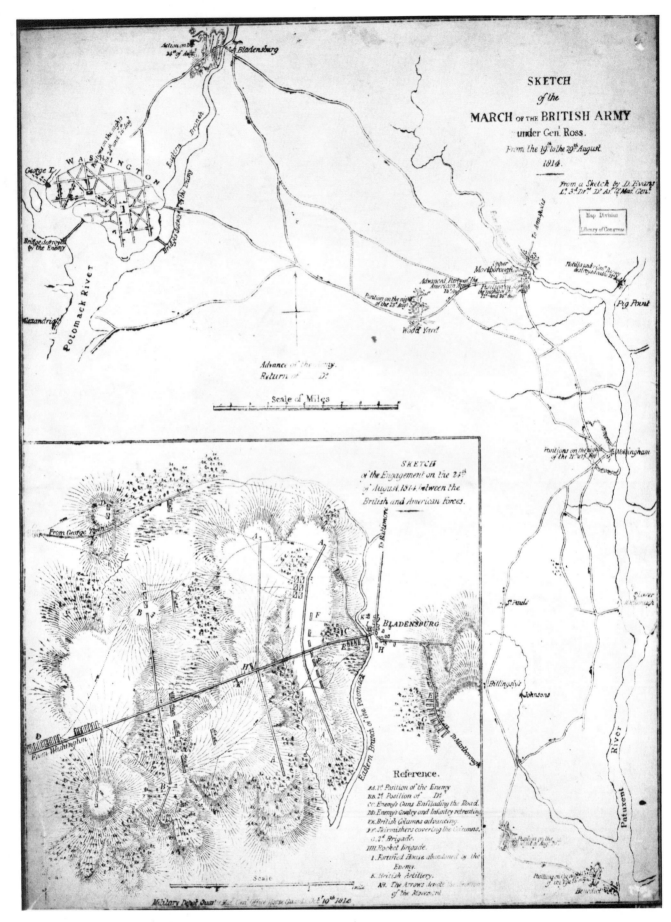

4.2 D. Evans, *Sketch of the March of the British Army under Genl. Ross,* London, 1814.
Courtesy of the Library of Congress

4.3 Title sheet from J. E. Woerl, *Das Koenigreich Wuerttemberg–Das Grossherzogthum Baden,* Freiburg, 1831–34.
Courtesy of the Library of Congress

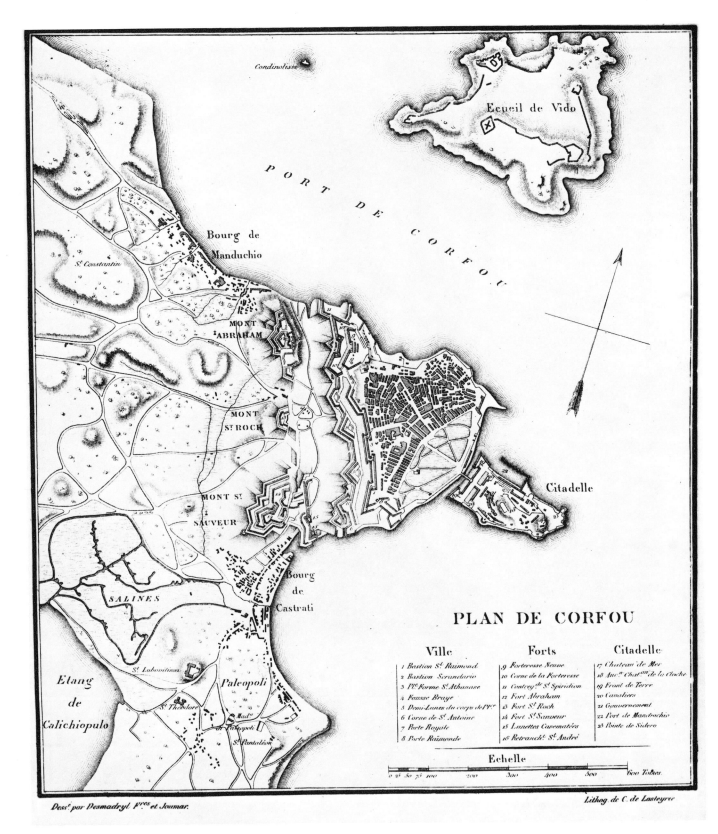

Condinolissa

Ecueil de Vido

PORT DE CORFOU

Bourg de Manduchio

St Constantin

MONT ABRAHAM

MONT St ROCH

MONT St SAUVEUR

Citadelle

SALINES

Bourg de Castrati

St Laboulissa

Etang de Calichiopulo

Paleopoli

St Theodore

Mad.

St Palæopoli

St Pantaléon

PLAN DE CORFOU

Ville	Forts	Citadelle
1 Bastion St Raimond	9 Forteresse Neuve	17 Chateau de Mer
2 Bastion Scrandario	10 Corne de la Forteresse	18 Anc.ne Chat.au de la Cloche
3 Pte Forme St Athanase	11 Contreg.de St Spiridion	19 Front de Terre
4 Fausse Braye	12 Fort Abraham	20 Canaliers
5 Demi-Lunes du corps de Pce	13 Fort St Roch	21 Gouvernement
6 Corne de St Antoine	14 Fort St Sauveur	22 Port de Manduchio
7 Porte Royale	15 Lunettes Casematées	23 Pointe de Sidero
8 Porte Raimonde	16 Retranch.t St André	

Echelle

0 25 50 75 100   200   300   400   500   600 Toises.

Dess.é par Desmadryl F.res et Joumar.

Lithog. de C. de Lasteyrie

4.4 "Plan de Corfu," in A. Schneider, *Nouvel Atlas pour servir a l'Histoire des Iles Ioniennes,*
Paris, 1823.
Courtesy of the Library of Congress

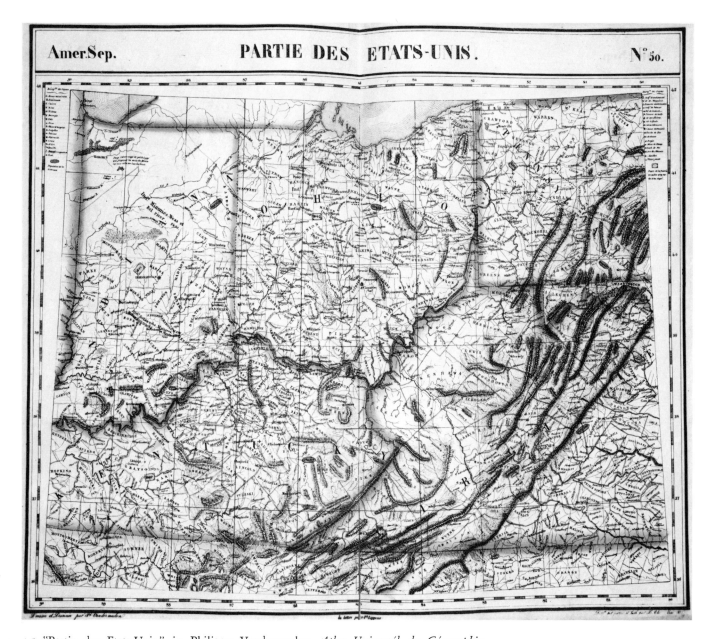

4.5 "Partie des Etats-Unis," in Philippe Vandermaelen, *Atlas Universél de Géographie,* Brussels, 1825–27, vol. 4.
Courtesy of the Library of Congress

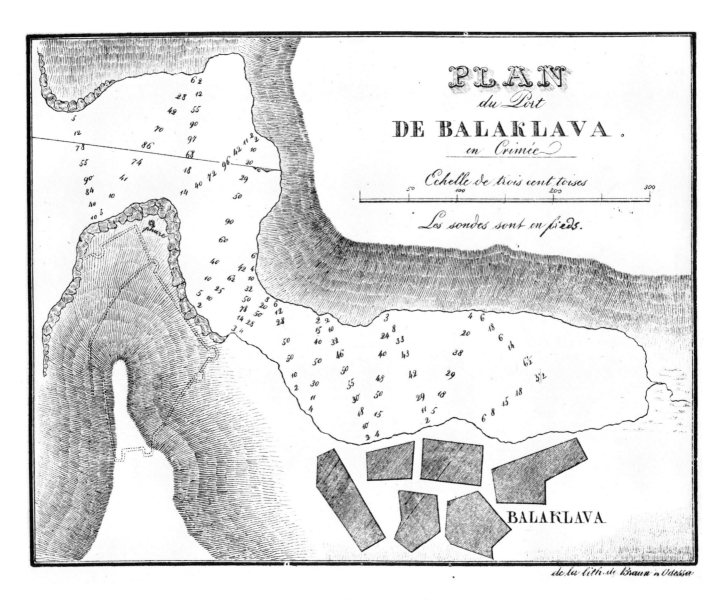

4.6 "Plan du Port de Balaklava," in E. Taitbout de Marigny, *Plans de golfes, baies, ports et rades de la Mer d'Azov,* Odessa, 1830.
Courtesy of the Library of Congress

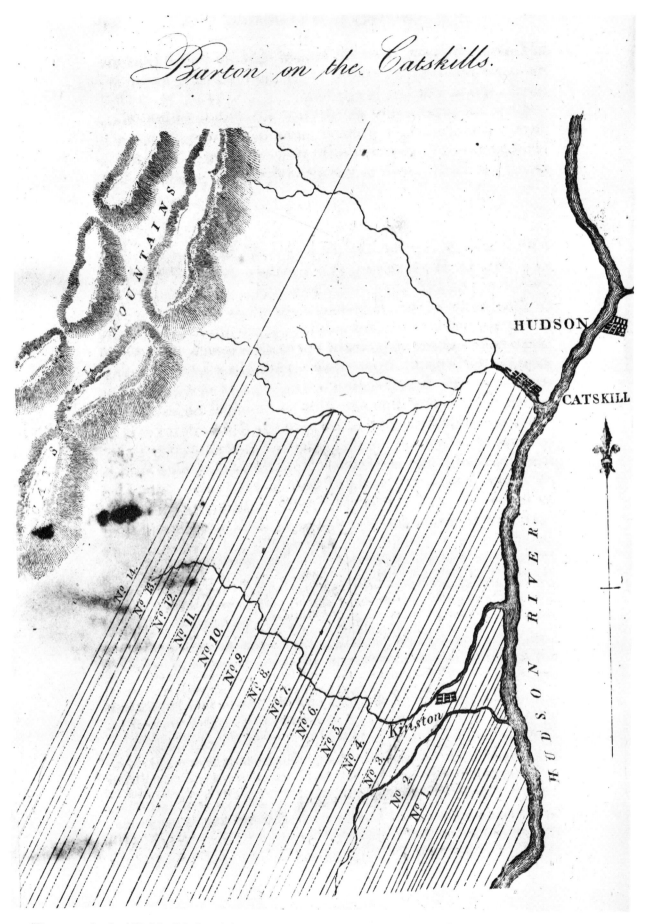

4.7 "Barton on the Catskills," in "Notice of the lithographic art," *American Journal of Science and Arts* 4 (1822): 169.
Courtesy of the Library of Congress

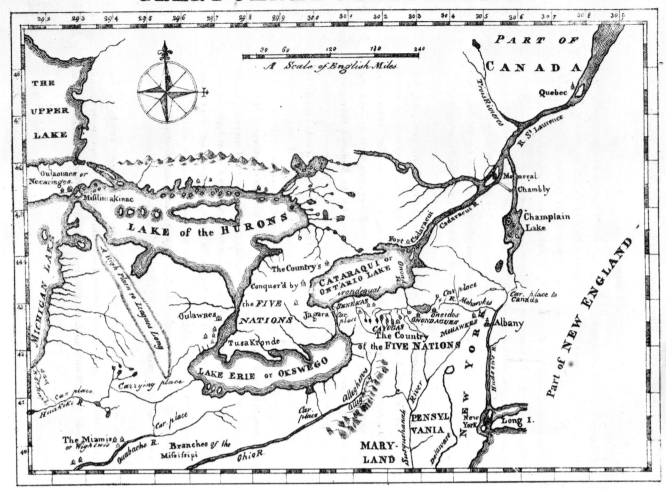

GRAND CANAL CELEBRATION.

Copy of a Map attached to Gov.ᵣ Colden's History of the FIVE INDIAN NATIONS

4.8 "Grand Canal Celebration," in Cadwallader D. Colden,
Memoir . . . presented . . . at the celebration of the completion of the New York canals, New
York, 1825

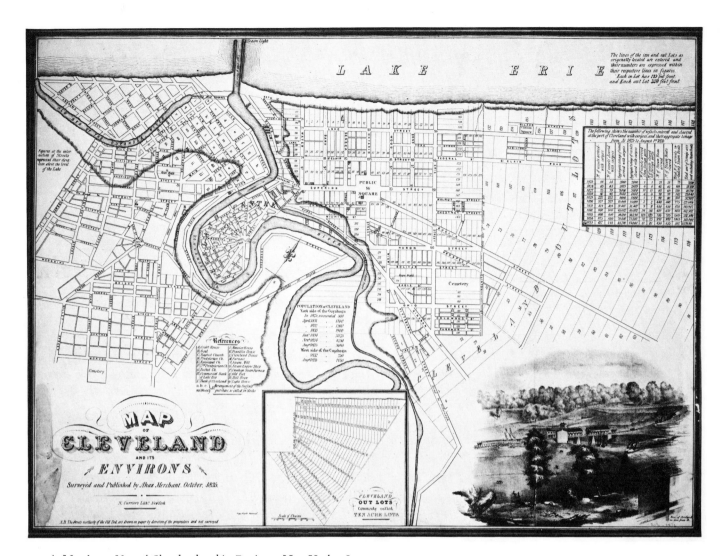

4.9 A. Merchant, *Map of Cleveland and its Environs,* New York, 1835.
Courtesy of the Library of Congress

turned to Munich. André remained, and in 1803 he published *Specimens of Poly-autography,* which described lithography. He continued to operate the lithographic press in London until 1805, primarily for printing artistic works. This continued to be its principal use after J. G. Vollweiler acquired the patent license. Vollweiler, however, also advertised letter and music printing.[13] Mumford observes that "this situation was in marked contrast to that obtaining in Germany, where the [litho-graphic] process was mainly applied to commercial ends including the printing of maps."[14]

Vollweiler's venture into lithography was, apparently, not a financial success. In 1807 he returned to Germany after selling the press and license to the Quarter-master General's Office, in the Horse Guards, Whitehall. D. Redman, who had served as pressman for both André and Vollweiler, was employed in the same ca-pacity, for a time, by the Quartermaster General's Office. "The story of the early days of lithographic printing at the Horse Guards from 1807," Mumford believes, "merits fuller investigation."[15] He reports that

the earliest map product of the press . . . is . . . in the British Museum Map Room in an album of what appears to be specimens of the mainly map work executed at the Horse Guards between 1808 and 1825. There is a plan of Bantry Bay dated 7 May 1808 which is not otherwise held by the British Museum.[16]

There are in the collections of the Library of Congress several of the maps printed at the Horse Guards' lithographic press. The earliest, published on 19 October 1814, is an uncolored "Sketch of the March of the British Army under Genl. Ross from the 19th to the 29th August, 1814" (fig. 4.2). On the same sheet is a "Sketch of the Engagement on the 24th of August 1814 between the British and American Forces" at Bladensburg, Maryland.

By the time Senefelder's . . . book appeared in England in 1819 . . . [Mumford notes] the press [in the Quartermaster General's Office] was undoubtedly well es-tablished, with a wide variety of maps and plans printed including items related to the Peninsular War as well as maps for the control of home forces in the British Isles.[17]

The *Gentleman's Magazine* reported in 1815 on the lithographic activity of the Quartermaster General's Office, noting that

His Majesty's Board of Ordnance have . . . erected a press at their Office, Whitehall, where . . . this mode of printing performed for them by Redman, was found very serviceable in giving immediate circulation to military plans of battles and other surveys, amongst the Staff-officers of the Army, the Ministers, and those connected with the Department of the Commander in Chief, during the late War.[18]

The account adds that Redman "resided nearly three years at Bath, during which . . . he printed a variety of drawings for Professors of the Art in that neighbourhood, and many more for Amateurs."[19] The Quartermaster General's operation was one of the earliest significant applications of lithography to military mapping. It is par-ticularly noteworthy because it preceded by several decades general adoption of the new process in England by commercial map publishers or by the British Ord-nance Survey.

Lithography was not soundly established in France before 1815, although a number of French scholars and printers investigated the new process soon after its invention. Frederic André, who was early associated with Senefelder, took out a French patent for lithography in 1802 and opened a small printing shop, which apparently did not survive. An 1807 André venture had greater success. In 1803, Edme-François Jomard, the distinguished orientalist and geographer, visited Regensburg, Germany, where he was introduced to the new process by Niedermayr, an early student of Senefelder. Jomard, who some years later became the first curator of maps in the Bibliothèque nationale and a founder of the Société de geographie, may have been the first professional geographer to become seriously concerned with lithography. In 1808 or 1809 André prepared lithographic plates for a report on archaeological studies Jomard had conducted in Egypt. Colonel Lomet of the French army visited Munich in 1806 to obtain lithographic stones and chalk and to receive briefing on the new technique. Lomet was convinced that lithography could be applied to printing maps and plans in France as it had been in Bavaria and Prussia. He was not at this early date, however, able to arouse much enthusiasm among his French colleagues.[20]

In the formative years of lithography, its most zealous promoters in France were Count Charles-Philibert de Lasteyrie-Dusaillant and Godefroy Engelmann. In 1810, Baron Aretin proposed to Lasteyrie that the two form an association. The baron offered to bring trained artists and lithographers to Paris. Count de Lasteyrie, who favorably considered the proposal, decided to visit Munich to become versant in the new art. He spent a month or more there, in 1812, studying with Clement Senefelder, brother of the inventor. Because of the war then in progress between France and Russia, German craftsmen were reluctant to move to Paris. Lasteyrie returned to Munich a year or two later for more concentrated and specialized instruction in lithography. He acquired presses, stones, and other essential lithographic equipment that were transported to Paris where, in 1815, he finally established a lithographic shop.

Godefroy Engelmann (1788–1839) is generally regarded as France's first successful lithographer. Elizabeth and Joseph Pennell believed that "no one contributed more than Engelmann to the development of lithography."[21] A native of Mülhaussen, in Alsace, Engelmann established his first lithographic press in that city in 1814. Two years later he opened a second shop in Paris, and within five years his was one of the leading presses in Europe. In 1815 Engelmann submitted samples of his work to the Société d'encouragement, Paris, of which Count de Lasteyrie was a founder. The December 1815 issue of the society's *Bulletin* carried a report by Lasteyrie about Engelmann's lithographs, which included a geographical map. In the report, the count related details of the establishment of lithography in France, including his own early efforts. He called attention to some of the difficulties encountered by pioneer lithographers and stressed the importance of using the utmost care at each step of the process. Engelmann's essay, submitted with his lithographic samples, was, in Lasteyrie's opinion, intended primarily to call attention to his own press.

Lasteyrie described Engelmann's samples as

quatre planches, dont la première représente un taureau au crayon avec des blancs de réserve sur un fond de bistre; la seconde, deux satyres faits à la manière en bois, et un sujet à la plume avec une teinte au bistre; la troisième, des notes de musique, et de l'écriture transportée du papier sur la pierre; la quatrième enfin, une carte géographique et une petite tête, l'une et l'autre gravées sur pierre.[22]

There was, regrettably, no description of the "carte géographique" or any indication of what area it portrayed. The report concluded that Engelmann's lithographs fell somewhat short of the quality achieved by German printers. Lasteyrie recommended, however, that the society commend Engelmann for the success he had achieved, acknowledge his essay, and express the wish that he continue the efforts that he had so competently begun.[23]

Engelmann did carry forward his lithographic work in France, and he was also instrumental in having presses established in Spain and Russia. In 1820, in collaboration with G. Berger, Engelmann published *Porte-feuille géographique et ethnographique*. Among the volume's numerous illustrations are thirty-two maps, all produced by lithography, and all but two of which are hand-colored. *Porte-feuille géographique,* a mathematical geographical textbook, may be one of the earliest teaching aids with lithographed maps.

French military applications of lithography lagged some years after such developments in England. Gräff reports that lithography was first applied to military functions in France about 1818, when a French general employed a lithographic press in Spain to reproduce orders and other official documents.[24] There is no specific mention of printing maps with the press. From 1817 on lithography flourished in France, with the main emphasis initially upon reproducing illustrations and works of art. Few lithographic maps published in France before 1820 are extant. Lithographic presses, reportedly, were also operating in 1820 in Austria, Belgium, Italy, the Netherlands, Spain, and Russia. There is meager record, however, of any cartographic items they may have printed in this period.

CARTOGRAPHIC LITHOGRAPHY,
1820–50

Publication of Senefelder's *Complete Course of Lithography,* in German, French, and English editions, in 1818 and 1819, established a watershed in the history and evolution of the art. Growth and development were uncertain and uncoordinated prior to 1820. The "early years of the 19th century," Fordham observed, "saw a good many experiments—in Germany bright boys in printing offices all over the country laboured earnestly and in secret to produce as show pieces unwieldy and enormous topographic maps . . . and there are on record at least half a dozen similar experiments in Germany, and four or five in France."[25] From 1820 on, the new printing process became more standardized and experienced rapid growth and development. This was true for the reproduction of maps as well as for other graphics. Nonetheless, copperplate engraving continued to be utilized in printing maps and plates for atlases, by both commercial and government cartographic publishers, well beyond the middle of the nineteenth century.

Expansion in cartographic production was stimulated by the several military campaigns that disrupted Europe during the first several decades of the nineteenth century. The governments of virtually every European country established official mapping departments during these years, usually under the direction of military general staffs. New techniques for surveying and mapping as well as improved instruments likewise contributed to the expansion. Not least of the stimulants to map production was the industrial revolution with its demands for better means of transport; improved roads, expansion of canal networks, and railroad construction, in turn, created new requirements for maps.

Guided by Senefelder's *Complete Course of Lithography,* and by lesser contributions by other practitioners, lithographic printing had parallel developments in the principal European countries during the eighteen twenties, thirties, and forties. A new technique introduced in one country was soon put into practice in other countries. Many of the applications anticipated by Senefelder were developed and perfected. Thus, zinc plates were substituted for unwieldy stone slabs as early as 1819, and some six years later the first cylinder power press was introduced. By the early 1830s limited success had been achieved in chromolithography, and the transfer method was in common use.

The lithographic process had become fairly well established by 1825. Engraving and etching on stone were still employed for certain detailed work, including some maps, but planographic printing had become the most common lithographic technique. Maps, and other images, were drawn or traced on stone, with pen and ink or lithographic chalk, or drawn, with lithographic ink, on specially prepared paper, from which they were transferred to stone or metal surfaces. The stone (or zinc) was washed with a weak solution of nitric acid, which accentuated the contrast between the inked and uninked portions, and increased the porosity and water-absorbing capacity of the latter. When a roller charged with greasy ink was passed over the stone, the ink was accepted by the drawn images and characters and rejected by the moistened areas. The final step in reproduction was to place a sheet of paper in a frame by means of which it was brought in contact with the inked surface. With application of pressure in a press, the image was transferred from stone to paper.

Development of a satisfactory press was one of the more vexing problems that confronted early lithographers. According to Twyman, many of them initially adapted presses that had been developed for letterpress and copperplate printing.[26] Not much, however, is known about early presses, and few examples have survived. Senefelder used an upright lever press, constructed of wood, which was known as a pole press. With two pressmen it had a reported capacity of one thousand impressions per day.[27] "Between about 1820 and 1825," Twyman relates,

the wooden star-wheel press became established in Germany, France, and England as the best kind of lithographic press for quality printing. Its principal features were: a bed with a tympan hinged to it, a cylinder across which the bed could be pulled, a winching cylinder which was linked by straps to the bed and which had a star-wheel fixed to its axle, a pressure system consisting of a lever and treadle linked by a connecting rod, and a scraper beam which hinged to the side of the press. Engelmann and Hullmandel are both known to have used presses of this kind.[28]

After 1830 metal replaced wood on most lithographic presses, but they continued to be manually operated until the middle of the nineteenth century. Twyman affirms that

right up to 1850 the hand press remained, if not unchallenged, at least unrivalled; for despite enthusiastic reports none of the powered lithographic machines patented before then seems to have been developed much beyond the experimental stage. After 1851 the hand press continued to be manufactured and used for transferring and proofing, and for printing editions of chalk drawings and similar work, but it no longer determined the development of commercial lithography, as it had done during the first half of the century.[29]

The range and scope of map printing expanded after 1820. Multiple-sheet topographic maps, nautical charts, educational maps, transportation maps, and atlases were added to cadastral and city plans, and general and military maps, which were among earlier cartographic works printed by lithography. Because they have assured the preservation of many early examples of lithographic cartography, atlases are of particular value in reconstructing the early history of the art.

LITHOGRAPHIC MAP PRINTING IN GERMANY, 1820–50

By 1834, the year Senefelder died, a number of lithographic presses were operating successfully in various German cities. Several other provinces followed the lead of Bavaria in establishing cadastral surveys, with plans printed by lithography. Cartographic products of private and commercial printers improved in quality and were issued in larger editions. Happily, many noteworthy examples for the period 1820 to 1850 are preserved in libraries and archives. City and town plans, still among the more common cartographic products, were more imaginative and better designed than those prepared during lithography's first two decades.

During the 1830s several medium-scale topographic map series, by Joseph E. Woerl, were lithographed at Herder's Kunst-Institut in Freiburg. Printed on twelve sheets, including an ornately embellished title sheet, is a 1:200,000-scale map of Württemberg, Baden, and Hohenzollern (fig. 4.3). The individual sheets are dated from 1831 to 1834. Engraving on stone was the technique employed. Relief features are shown with considerable detail, and cities and an extensive network of roads are overprinted in red.

The suggestion for two-color map printing was proposed to Herder, in the early twenties, by Johann Heinrich Weiss. There were numerous problems to surmount, and progress was slow. In 1828, the project was taken over by Weiss's son-in-law, Woerl, who carried it through to successful accomplishment. Krause indicates that Herder published Woerl's nineteen-sheet *Topographische Karte des Rheinstrom's,* at the scale of 1:20,000, in 1828 and his twenty-five-sheet *Atlas von Europa* in 1829.[30] Neither of these publications has been examined, however, and it is not certain whether or not they were printed in two colors. The Woerl-Herder publications are believed to be the first successful applications of two-color lithographic printing to maps.

An early lithographically printed official topographic map series is *Karte von dem Grossherzogthume Hessen* (grand duchy of Hessen), at the scale of 1:50,000, which was published by the Hessischen Generalstabe. All but one of the thirty sheets were lithographed by C. Kling. Several sheets have manually applied boundary colors. The series, which is undated, was probably published after 1840. A similar map, at the same scale, of Kurfürstenthum Hessen (electorate of Hessen), was published by the General Staff of the electorate. The forty sheets, which were issued between 1840 and 1857, are credited to a number of lithographers.

The potential for producing inexpensive maps by lithography for teaching purposes was early recognized. William Engelmann's *Bibliotheca Geographica,* published in 1858, for example, lists a number of school atlases, globes, and wall maps that were lithographically reproduced from 1819 onward.[31] Not until the 1840s, however, was lithography extensively employed in Germany in printing school atlases.

The earliest German lithographic atlas in the collections of the Library of Congress is Wilhelm E. G. Schlieben's three-volume *Atlas von Europa nebst den Kolonien.* The first volume was published in Leipzig in 1825, and the other two volumes were issued between 1825 and 1830. A number of maps in all volumes carry the inscription "O. Herrman, lith." The *Atlas der Alten Welt in XVI illuminirten Charten* was published at Düsseldorf, in 1829, by Arnz and Company. This historical school atlas includes sixteen maps, all of which are hand-colored. The *Atlas des Königreichs Preussen,* published in Erfurt by Müller's bookshop in 1831, was lithographed and printed at C. A. Eyraud's Neuhaldensleben shop. Administrative borders are hand-colored. There are six uncolored lithographic maps, a number of profiles, and several pages of text and tables in the *Atlas zur Expedition der Franzosen und Engländer gegen der Citadelle von Antwerpen im Jahre 1832,* published in Berlin in 1833 by E. S. Mittler. Several of the maps in this military atlas are hand-colored. Although, as is evident from these examples, lithography was applied to atlas printing by the end of the third decade of the century, the major German reference atlases continued to rely upon engraving for another half century. *Stieler's Handatlas,* which was introduced in 1817–23, did not shift to lithography until late in the nineteenth century. Justus Perthes, publisher of the *Handatlas,* did print school wall maps by lithography as early as 1838.

Lithographic printing advanced more rapidly in France after 1820 than in Germany, and French lithographers contributed much toward perfecting the art. The presses of Engelmann and Count de Lasteyrie were most successful, but there were a number of other lesser shops in Paris as well as in other French cities. "For 20 years," Pennell wrote, "lithography in France was so popular, its practice so widespread, and its results so splendid that it is difficult to give a complete record."[32] Although the main emphasis, between 1820 and 1840, was on reproducing works of art, noteworthy progress was also made in printing maps.

Cartographic lithography received strong support from a report on the question "Si la Lithographie peut-etre appliquée avec avantage à la publication des Cartes Géographiques," prepared by a commission of the Société de geographie. The report, published in vol. 5 (1826) of the society's *Bulletin,* was signed by Jomard, who served as rapporteur. The five distinguished members of the commission carefully weighed the respective advantages and disadvantages of engraving and lithography in map reproduction. Specifically, they sought "to determine whether lithography presents significant and adequate advantages, when applied to topographic presentation, and to geographic maps; if it has utility in executing boundaries, mountains, waters, and forests, of cultural variations and terrain features, or for the inscription of names and lettering of the type used on maps."[33]

In the course of their deliberations the commision members examined lithographic maps and plans of various sizes and scales. The report commended Engelmann, Desmadryl, and the firm of Cosnier and Renou for their good work in cartographic lithography. That lithographic map printing was well established by 1826 is evident from the observation that "it would be difficult to enumerate all the examples of this type of lithography."[34] The commission believed that lithography was super-

FRENCH LITHOGRAPHIC
CARTOGRAPHY, 1820–50

ior for more than half of the qualities examined. "The final conclusion then," the report states,

is to the advantage of lithography, on the basis of cost; we hasten to add that copper engraving has, and will undoubtedly retain for a long time, a real superiority over lithography as an artistic medium, and that it alone is able to create those topographic masterpieces which bring honor to French artists. It is a credit to lithography that it has approached this goal.[35]

The commission members hoped lithography might provide the medium for producing maps for teaching geography. They suggested that deficiencies in geographical knowledge in the past were definitely related to the paucity of maps. They were moved, therefore, to give every encouragement to lithography, which support they expressed with vehemence.[36]

The commission was not, however, disposed to abandon copper engraving. The two arts, the members declared, were not exclusive.

Each has its purpose. The older and more exact one will continue to be applied to geographic maps, to works of large dimension, to map series or atlases which demand a high degree of uniformity, and finally to maps which may be reprinted over a long period of time. On the other hand, the new art will be applied to geographic studies, to individual maps, to the cartographic requirements of travel and commerce; lithography's advantages respond to these most urgent needs. Meanwhile the first [i.e., engraving] will hold over the second two significant advantages: (1) the engraved plates can be retained indefinitely, without alteration; (2) the option, at any time, to make desired corrections as new information comes to hand.[37]

The commission's encouragement to lithography as an effective medium for producing educational and special purpose maps unquestionably served to stimulate geographical studies and fostered the development and growth of thematic cartography. Prior to the Société de geographie study, however, some few school maps had been published in France. One early example in the collections of the Library of Congress, entitled *Mappemonde dressée pour l'usage des Colléges, 1821,* was printed by "Selves fils, lithographe de l'Université." The world is shown on a double-hemisphere projection, with crude "caterpillar" hatching to indicate the principal mountain ranges. Country boundaries are not shown.

Of the several lithographic cartographic items noted in the report of the Société de geographie's commission, we may cite Baron Antoine Schneider's *Nouvel Atlas pour servir a l'Histoire des Iles Ioniennes,* published in 1823 by Dondey-Dupré, Père et Fils, Paris. It predates by several years Vandermaelen's world atlas, which is often cited as the first lithographically produced atlas. The *Nouvel Atlas* includes several illustrations, a number of statistical tables, and twelve neatly executed maps and plans (fig. 4.4). The latter all bear the notation "Lithog. by C. de Lasteyrie."

Several French lithographers considered the problem of printing in color as early as 1820 and, in 1828, the Société d'encouragement proposed a prize of 2,000 francs for the invention, and successful application, of a process for color printing. In December 1836, Engelmann submitted, to the Société industrielle of Mulhausen, some crayon lithographs printed in color, regarding which a favorable report was made on 29 March 1837.[38] At the 1839 Industrial Exposition a large number of examples of color printing were displayed by Engelmann Sons, Lessore, Quinet, Carles, and Garson, of Paris, Simon Sons, of Strasbourg, and Barbat, of Chalons-sur-Marne.[39]

In 1843, the Kaeppelin lithographic firm of Paris printed a geologic map of the Paris Basin, which included eleven different colors, obtained in four printings by superposition of colored inks. The map, entitled *Carte Geognostique du Plateau Tertiaire Parisien,* was compiled by Victor Raulin, vice secretary of the Société geologique de France. It is at the scale of 1:300,000 and measures 28 × 35 inches. Some twenty-two months later France's Imprimerie royale, under the direction of chief lithographer M. Derenémesnil, printed, in twenty-three colors, *Feuille d'Assemblage de la Carte Geologique de la France.* In 1845, *Le Lithographe, Journal des Artistes et des Imprimeurs* (Paris and Rotterdam) published a lengthy article on the subject of "Chromolithographie." As of 1845, some eighty Paris lithographic presses were, reportedly, printing in color.

France's rapidly developing railroad system provided a market for inexpensive maps, which lithographers were quick to exploit. Among particularly interesting early railroad maps is *Carte du Chemin de Fer de Strasbourg à Bâle,* at the scale of 1:80,000. This uncolored strip-type map, which measures 5 × 66 inches, was published in 1841 by the Strasbourg lithographic firm of Levrault.

Lithography was early viewed as a potential medium for preparing reproductions of unique or rare paintings, portraits, and maps. An early cartographic facsimile is João de Castro's *Roteiro em que se contem a viagem que fizeram os Portuguezes no anno de 1541, partindo da nobre cidade de Goa atee Soez . . . ,* published in Paris in 1833. Bound within the volume are sixteen folded maps printed from lithographic stones. The reproductions were made from hand-drafted tracings or drawings of original manuscript maps preserved in the British Museum. The maps and charts are embellished with pictorial renderings of ships, contingents of mounted troops, vegetation types, native inhabitants, and scenic profiles. Maps in the Library of Congress's copy of de Castro's book are uncolored. In the library's Geography and Map Division there is a second set of the maps, laid flat and bound in hard covers. All but one of the maps in this "atlas" are colored.

Between 1840 and 1860 several ambitious facsimile atlases were published, primarily with lithographic reproductions. In this prephotographic era, the reproductions were based on hand tracings or drawings. Manuel Francisco de Barros, second viscount de Santarém, a Portuguese gentleman and historian, published, in 1840–41, Portuguese and French editions of a scholarly study entitled *Recherches sur la priorité de la découverte des pays situés sur la côte occidentale d'Afrique, au-delà du Cap Bojador, et sur les progrès de la science géographique, après les navigations des Portugais, au XVᵉ siècle.* The volume was accompanied by an atlas with twenty-one plates on which were reproduced twenty-three cartographic "monuments." The book and maps were intended to support the thesis of Portuguese priority in the discovery of the west coast of Africa.

In 1842 Santarém issued a second edition of the atlas, which was expanded to include reproductions of a number of medieval *mappae mundi.* The compiler continued to select for reproduction additional maps from various European repositories, and a third, greatly augmented, atlas was published in 1849. The title is *Atlas composé de mappemondes, de portulans et de cartes hydrographiques et historiques depuis le VIᵉ jusqu'au XVIIᵉ siecle. . . .* Because the atlas was in loose-leaf format it is difficult to determine the total number of maps issued. The copy in the Library of Congress includes reproductions of seventy-nine world maps and globes.[40]

From 1834 until his death, in 1856, Santarém lived in exile in Paris. Nonetheless, the Portuguese government provided financial assistance for his research and published his atlases. The latter were all printed in Paris. Seven of the seventy-nine maps in the Library of Congress's Santarém atlas were reproduced from copper-engraved plates, all others were lithographically printed. Certain of the maps were engraved, rather than drawn, on stone, as indicated by the note, "Gravée sur pierre," which appears on some sheets. Most of the facsimiles were printed at the Kaeppelin or Lemercier presses in Paris, with lithographic drawings prepared by J. Feuquières, L. Bouffard, or Schwaerz. There are two hand-colored maps, one drawn by Bouffard and printed by Lemercier, the other the work of Wursler & Cie., of Winterthur, Switzerland. This is the latter firm's only contribution to the atlas.

In an official report, submitted to the Portuguese Foreign Office on 5 June 1854, Santarém explained why lithography was selected as the major reproduction medium. "Before beginning the publication," he wrote,

another subject worried me: to find out whether the monuments should be engraved on copper or lithographed. Although I had made up my mind on the subject, I convoked five of the best engravers in Paris; after the two methods were discussed, it was agreed that if copper engraving has the advantage of more perfection in the drawing and in the lines of wind-roses than the lithographic stone, lithography was preferable, because the tracing or drawing of the artist could be transferred in facsimile better on the stone, and reproduce the design of the geographical monument better than on copper which, however perfect the engraving might be, reproduced a copy, besides the excessive cost of the latter method. They agreed to this, adducing various technical reasons. . . . These difficulties settled, I started immediately the engraving of the first monuments to illustrate the text of the *Recherches,* and such was my eagerness in this matter that on 14 June 1841 I already had fourteen engraved and printed, and by 18 February 1842 I had sixteen more engraved, so that when the *Recherches* were published they were accompanied by the *Atlas* containing thirty monuments. . . .[41]

In the *avertissement,* which accompanies his facsimiles, Santarém wrote, "We believe . . . that we can say with full confidence that we have spared no pains in producing this publication, *the first of its kind* that has been made, to give it the faithful and accurate reproductions of the actual monuments of the geography of the Middle Age." Santarém's claim to primacy in publishing a systematic series of facsimile maps was disputed by Edme François Jomard who, in 1828, had been placed in charge of the newly established cartographic department in France's national library. Jomard claimed that, shortly after assuming the position, he had conceived the idea of preparing a facsimile atlas by reproducing selected historical maps from the Bibliothèque's collections. Because of other obligations the project remained inactive for a decade or more. When the first of Santarém's atlases was published, Jomard, as stated in the *note preliminaire* to his atlas, "decided to publish those plates already completed." The final facsimiles in Jomard's series were not issued until 1862, shortly after the demise of the compiler. The title page for the collection reads, *Les Monuments de la géographie ou recueil d'anciennes cartes européennes et orientales . . . contenant des recherches pour servir a l'histoire des découvertes et des sciences géographiques.* Jomard's *Monuments* includes twenty-one world maps on eighty-one plates. All were reproduced lithographically, with most of the drawings by Eugen Rembielenski and printing by Kaeppelin & Cie. There is one hand-colored facsimile.

In the history of cartography it is immaterial whether credit for the first general facsimile atlas is granted to Jomard or to Santarém. By making knowledgeable selections of early maps both contributed to the establishment of cartographic history as a systematic field of study. Moreover, employing the new lithographic technique to reproduce noteworthy historical maps, they made these monuments of cartography available for study by scholars and researchers throughout the world, and advanced the development of lithographic cartography.

In view of the study on the application of lithography to map printing, made under the sponsorship of the Société de geographie, it is interesting to note the early inclusion of lithographic maps in the society's *Bulletin.* Although lithographic illustrations appeared in the journal as early as 1825, it was not until the 1828 edition that two maps, printed with the new technique, appeared. From this date on lithographic maps were included in various editions of the *Bulletin,* although engraving likewise continued to be employed for some map illustrations well into the thirties.

In 1830 Lieutenant-general Jean Jacques Pelet became the head of France's Dépôt de la guerre. He displayed an early interest in lithography, and beginning in 1838 the Dépôt employed the new technique to print département maps that were derived from the 1:80,000-scale map series. Twenty départements were mapped under this program between 1839 and 1872. The maps carry this statement:

Département de ———, extrait de la carte topographique de la France, levée par les officiers d'état-major et gravée au Dépôt général de la Guerre sous la direction du lieutenant-général Pelet, publié avec l'autorisation du Ministère de la Guerre. Paris 18– –, autographié et imprimé chez Koeppelin et Cie., 15, quai Voltaire.[42]

The Dépôt also used lithography to reproduce maps for the eight-part atlas that accompanied François Eugene de Vault's *Mémoires militaires relatifs à la succession d'Espagne sous Louis IV. . . .* This impressive work, published between 1835 and 1862, has an introduction by General Pelet. Prepared under the direction of Pelet, and published by the Dépôt général de la guerre in 1844, was the *Atlas des Campagnes de l'Empereur Napoleon en Allemagne et en France.* It consists of eleven folded maps and a title page in a slipcase. The title page and eight of the maps are engraved. Two maps were lithographed by Kaeppelin and one by Herder.

The development of lithography in England closely paralleled that in France. It was through the efforts of Rudolph Ackerman that an English edition of Senefelder's *Complete Course of Lithography* was published in 1819. The growth of the art in England during the next several decades was largely related to the efforts of Ackerman and Charles Hullmandel, an artist.[43] In 1824 Hullmandel published *The Art of Drawing on Stone,* which was reissued in 1833. Among the illustrations in both editions is a lithographic map of Vieux-Brisach in Germany. Notwithstanding the early application of the technique to cartographic printing by the Quartermaster Department, use of lithography for printing maps lagged in England. Darlington and Howgego, in the introduction to their *Printed Maps of London, circa 1553–1850,*

LITHOGRAPHIC CARTOGRAPHY IN ENGLAND, 1820–50

for example, note that "lithography was little used by publishers of London maps until almost the end of the period covered by the catalogue."[44]

Both the British Ordnance Survey and the Admiralty continued to engrave their maps and charts on copper for another half century or more. Darlington and Howgego observe that "some of the sheets of the large-scale skeleton Ordnance map [of London] were also reprinted [ca. 1854] by lithography; they appear to be some of the earliest maps for which this process was used by the Ordnance Survey."[45] Around 1826, the Secretary of the Admiralty, J. W. Croker, related to Hydrographer W. E. Parry the merits of lithography. He observed that "though not so neat as engraving," the new technique had the virtues of "cheapness, celerity, and ease of working, particularly for maritime charts [that are] not so crowded as geographical maps."[46] Parry questioned the advantages of lithography, and he was further deterred by the problem of making corrections on stone and by the potential risk of damage to, or destruction of, the stones. He concluded that it was not yet the time to shift from engraving to lithography.

A number of English commercial map publishers likewise continued to print maps from copper plates until the second half of the nineteenth century. Thus, W. & A. K. Johnston, founded in Edinburgh in 1825, did not shift to lithography until 1855.[47] By 1860 the company had installed a rotary lithographic press, and chromolithography was introduced around 1865. Similarly, John Bartholomew & Son, Ltd., also of Edinburgh, did not adopt lithography for map printing until 1880, two decades after the firm was established. George Philip & Son, of London, which was founded in 1834, did install lithographic presses in 1846, although they continued to print some maps from engraved plates after this date.[48]

In the specialized field of railroad mapping, lithography did gain a foothold in the middle and late 1830s. The first appearance of English railroad lines on a map, however, was in a foreign publication, Philippe Vandermaelen's *Atlas de l'Europe,* published in Brussels in two volumes between 1829 and 1833. Of particular interest are the several atlases prepared to accompany reports of the Irish Railway Commission. The first is titled *Plans of the several lines laid out under the direction of the Commissioners 1837. Presented to both Houses of Parliament by command of Her Majesty.* It includes two parts, with some folded maps, bound in one folio volume. Part one, prepared by Civil Engineer Charles Vignoles, has eleven map plates, all lithographed by R. Cartwright of London. Part two, prepared under the supervision of John MacNeill, Civil Engineer, has five map plates and an index map, all reproduced from copperplates engraved by J. and C. Walker. A general plan of Dublin showing the course of the Railway Colonnade, an architectural drawing of the Colonnade, and a map locating railway surveys through North Wales, prepared by Vignoles, are also bound in the volume.

C. F. Cheffins, lithographer, London, published in 1838 a map of the *London and Birmingham Railroad* and of the *Grand Junction Railway and Adjacent Country.* In the same year J. R. Jobbins, also of London, printed, by lithography, a *Map of the London and Southampton Railway,* and *Railway Map of England and Wales.* The latter was published under the imprint of W. A. Robertson. By 1840 lithographically printed railroad maps were quite common in the British Isles.

In its third year of publication, 1833, the *Journal of the Royal Geographical Society* included hand-colored lithographic maps. In subsequent issues, during the next

quarter-century or more, engraving and lithography shared responsibility for cartographic illustrations.

Benjamin P. Wilme published, in 1846, *A hand book for plain and ornamental
mapping*. In the introduction he noted that "the increased demand for maps has
of late years been so great as to produce an enormous influx of engineers, surveyors,
and draughtsmen."[49] The author gives a possible explanation for the reluctance
of some British official and commercial publishers to print maps with the new
technique. "Lithography," he wrote,

has of late years become a very favourite medium with Engineers and Surveyors, for
the production of duplicate plans for Parliamentary deposit. . . . It is to be regretted
that this art, (as applied to the purpose above named,) almost essential to the surveyor, has fallen into the hands of an ignorant class of persons, viz., the picture-
copiers and lithographic printers. It were impossible to detail the mischief annually
done by persons being intrusted with this class of business, who are totally ignorant
of the construction or use of maps. . . . The remedy is simple, ... Let surveyors
lithograph their own plans, or employ their draughtsmen upon them; much time
[Wilme believed] would thus be saved.[50]

Although the booklet gave detailed "instructions for lithographing plans, sections,
and drawings," it is unlikely that many surveyors attempted to print their own maps.
Many English mapmakers, official and private, continued to reproduce maps by engraving until the new technique had been perfected to embrace zincography, color
printing, transfer, power presses, and photo lithography. In a paper read before the
London Society of Arts in December, 1847, William Smith Williams observed that
while "lithography was rapidly making its way in Germany and France, it was
scarcely known in England, and was regarded as an inferior kind of engraving." He
offered the explanation that the "British were not so well versed in chemistry or
skilled in drawing."[51] At this date (i.e., 1847), Williams himself still believed that
"lithography is not to supplant engraving, but to supply a desideratum in that art;—
the multiplication of original drawings or sketches."[52]

An interesting inflatable paper globe was printed by lithography in London,
around 1835. The patentee was C. Pocock, of Bristol, who is described, in the cartouche, as "Author of various scientific Spheres, stationery and revolving." Ebe
Pocock is credited as "Del." (Delineator) and W. Day as "Lithog." The hand-
colored globe, when inflated, has a diameter of approximately six feet.

Ian Mumford notes that "in the 1830s the increasing interest in Parliamentary
activities resulted in heavy demands for quick response printing. Hansard was using
lithographed maps in parliamentary papers by 1840, and before long Day became
lithographic printer to the Queen."[53] He adds that "in his early partnership with
Haghe, Day is stated to have printed a plan of Westminster from zinc in 1835."

Anastatic printing, a variation of lithography, invited some attention in England
during the latter years of the 1840s. The process was invented around 1840 by
Charles Frederick Baldamus of Erfurt, Germany. Joseph Woods took out English
patents in 1844 on Baldamus's invention, for which he proposed the name *anastatic
printing*. The February 1845 issue of *Art-Union* gives the following description of
the process:

Let us suppose a newspaper about to be reprinted by this means. The sheet is first
moistened with dilute acid and placed between sheets of blotting paper, in order that

the superfluous moisture may be absorbed. The ink neutralizes the acid, which is pressed out from the blank spaces only, and etches them away. . . . The paper is then carefully placed upon the plate with which the letter-press to be transferred is in immediate contact, and the whole passed under a press, on removal from which, and on carefully disengaging the paper, the letters are found in reverse on the plate, which is then rubbed with a preparation of gum, after which the letters receive an addition of ink, which is immediately incorporated with that by which they are already formed. These operations are effected in a few minutes. The surface of the [zinc] plate round the letters is bitten in a very slight degree by the acid, and on the application of the ink it is rejected by the zinc, and received only by the letters which are charged with the ink by the common roller used in hand-printing. Each letter comes from the press as clear as if it had been imprinted by type metal; and the copies are facsimiles which cannot be distinguished from the original sheet.

It is evident from this that anastatic printing was originally looked to as a means of facsimile reproduction of previously printed letterpress, illustrations, or maps. It was soon learned, however, that new drawings, prepared with proper ink, could also be transferred to the zinc plate. During the period of its greatest popularity several maps were reproduced by the process at Cowell's Anastatic Press, London, in the early 1850s. Peter Brutt's *Plan and section showing proposed works for the Harwich Docks, Pier and reclamation from the River Stour* was published in 1852, and a *Military Sketch-map of England and Wales in Militia Brigade Districts shewing the primary hill and railway systems with Militia Depots,* and a *Plan of an Artillery Defence of East Suffolk to accompany a Memoir,* illustrated R. A. Shafto Adair's *The Militia of the United Kingdom,* published in London in 1855. Anastatic printing also was used for certain specialized map reproduction jobs at the British Ordnance Survey between 1850 and 1870.

Notwithstanding the enthusiastic accounts about the process in professional journals, and the support from such prominent individuals as Michael Faraday, anastatic printing did not achieve the success hoped for by its inventors. In 1862, in a volume on the *History of Printing,* it was reported that

the anastatic process of lithography, which caused considerable sensation when first invented, was expected to prove a most valuable extension of the art, since it was supposed to combine the advantages of lithographic drawing or transfer, with ordinary letter-press printing. . . . In practice . . . the plan was not found to answer as to be remunerative, and for the present it appears to have fallen in abeyance.[54]

LITHOGRAPHIC MAP PRINTING IN BELGIUM, THE NETHERLANDS, AUSTRIA, AND RUSSIA, 1820–50

Lithography was early introduced into the Low Countries, and a few maps were produced with the new technique before 1820. There were plans to engrave, on stone, sheets of the first topographic map of the Netherlands, based on surveys carried out between 1815 and 1830. A lithographic press was installed at Military Survey headquarters in Ghent in 1827. Because of the revolution of 1830, which led to the separation of Belgium from the Netherlands, the topographic map was not printed at this time. Prior to the anticipated printing of the map, the lithographic press at the Military Survey reproduced ten maps for the *Atlas der Kaarten en profillen, behoorende tot onderzock der beste rivierafleedingen,* by J. E. van Gorkum, 1827–28. This river survey includes the first official maps of the Netherlands printed lithographically.[55]

Cartographic lithography in Belgium experienced its greatest growth after 1825, guided and nurtured by Philippe Vandermaelen. Born in 1795 of well-to-do parents, Vandermaelen received his early education at Brussels's Lycée imperial. He spent about ten years in commercial pursuits, then studied mathematics under a Professor Pagini, a refugee scholar from Italy. This provided background for a career in map publishing upon which he embarked in 1824. His first project was an ambitious one, to publish an atlas for the world showing all parts at the uniform scale of 1:1,641,836. Vandermaelen personally compiled or drafted a number of the plates. The first part of the *Atlas Universel de Géographie* was published in 1825, and all six volumes were completed by 1827. It was dedicated to William I, king of the Netherlands. The atlas was distinctive in being printed by lithography under the direction of Henry Ode, the first world atlas to be so produced (fig. 4.5). The *Atlas Universel* sold for approximately $800 a set and, despite the high cost, more than eight hundred copies were sold.[56]

The success of his first cartographic effort encouraged Vandermaelen to publish the *Atlas de l'Europe, a l'échelle de 1:600,000 . . . dressé sur des matériaux rassemblés et les cartes construites par les célèbres géographes.* The two-volume work, dedicated to the prince of Orange-Nassau, was issued between 1829 and 1833. It was reprinted in 1840 in a limited edition of one hundred copies. The *Atlas de l'Europe,* which includes one hundred plates, was designed by Henri Perkin and lithographed under the direction of Pierre Doms, with geographical locations by F. Charles and relief representation by L. Bulens. All four specialists were trained in the Établissement géographique de Bruxelles, which Vandermaelen founded in 1829. Compilation and printing were carried out in the offices of the Établissement. The lithographic technique employed for the atlas may have utilized zinc plates. The process, perfected by Vandermaelen and his associates, "combined the rough-sketch in acid with the completed work in dry-point etching" and produced "results comparable to the best copper plate work."[57]

Because of its high price, and political and economic disruptions that resulted from the Belgian Revolution of 1830, the *Atlas de l'Europe* was not as financially successful as Vandermaelen's earlier effort. Nonetheless, the Établissement géographique continued an active program of cartographic publication for more than forty years. Other noteworthy products included a lithographic reproduction of Ferraris's *Carte Marchande* of Belgium, at the scale of 1:86,400 (1831), the *Topographic Map of Belgium,* in twenty-five sheets, at the scale of 1:80,000 (1841–53), a two hundred fifty-sheet *Topographic Map of Belgium,* at the scale of 1:20,000 (1846–54), cadastral surveys at scales up to 1:2500, André Dumont's geological map of Belgium in nine sheets (1857), the first of its type for that country, a railroad atlas of Belgium in fifteen plates (1840), and a number of city plans and province maps.

Vandermaelen's Établissement géographique had a tremendous impact on map publishing in Belgium and in Europe during the middle half of the nineteenth century. However, by the time its founder died, in 1869, improved reproduction techniques such as photolithography, color printing, and the rotary steam press had superseded the lithographic processes employed in the Établissement. Official assumption of surveying and mapping responsibilities likewise heralded the decline of such private cartographic operations. Vandermaelen's son kept the workshop open until 1878, but the later years were not particularly productive.

The Austria-Hungarian Empire in 1818 established a cadastral survey, for which the plans were printed by lithography. The same year the Quartermaster Department of the Military General Staff set up a topographical-lithographic printing plant.[58] A great number of maps were reproduced lithographically in Austria-Hungary between 1826 and 1850, by both official and commercial publishers. Among the output of the Quartermaster General's plant were a road map of Bohemia and Moravia, at the scale of 1:432,000, a nine-sheet road map of the empire, at the scale of 1:864,000, a 1:400,000-scale map of Russia, in sixteen sheets, and environs maps of Vienna, Graz, Brunn, and Lemberg, at the scale of 1:14,400. Nischer von Falkenhof states that with the last listed city plans, published in 1828, their compiler, Franz Ritter von Hauslab, "introduced color printing to lithography."[59]

Between 1845 and 1847 the Austria-Hungarian Military Geographical Institute reproduced, by lithography, *Generalkarte von Europa,* at the scale of 1:2,592,000, which included twenty-five sheets.[60] An impressive example of Austrian commercial lithographic cartography is *Grundriss der Haupt u. Residenz Stadt Wien,* published in 1832, which is 42 x 46 inches in size. It was published by Antony, Freiherrn von Guldenstern.

Russian lithographic mapping developed slowly during the first half of the nineteenth century. In 1848, Philippe Vandermaelen declined an invitation to assume management, in Saint Petersburg, of the Russian Imperial Mapping Office.[61] A distinctive example of early lithographic cartography in Russia is E. Taitbout de Marigny's *Plans de golfes, baies, ports et rades de la Mer d'Azov* (fig. 4.6). It was printed at the lithographic shop of Alexandre Braun, and published at Odessa in 1830. The thirty-one chart plates and two pages of profiles were all produced by lithography. The detail and neatness of hachuring, coastline delimitations, and soundings approach the quality of engraved charts of the period.

LITHOGRAPHIC CARTOGRAPHY
IN THE UNITED STATES

The foundations for lithography were not laid in the United States until the incunable period of the new art had run its course in Europe. Bass Otis is generally credited with having introduced the new technique. Silver relates that "using a stone from Munich which had been presented to the American Philosophical Society by Thomas Dobson, Bass Otis drew a lithograph in Philadelphia for the July, 1819, issue of the *Analectic Magazine.*"[62] Otis's contribution was purely experimental, and his interest in lithography apparently was ephemeral.

American dependence upon European lithography and lithographers continued for three or four decades. Marzio states that "not one American contributed a new important idea for the design of presses or the use of stones before the Civil War."[63] He adds that "the thousands of lithographers who drew and printed lithographs in the United States used European manuals, Bavarian stones, and presses of English, French, or German design."[64] "The standardization of lithography at this early date," Marzio records,

meant that the entire craft was imported to the United States. Americans received it enthusiastically, honoring it as a fixed science with a complete set of laws. Rather than attempting to improve the process, their initial reaction was to look for local stones that worked as well as the magical specimens from Solenhofen.[65]

As might be anticipated, American engravers were among the first to be interested in lithography. In few instances, however, did an engraver wholly abandon copper in favor of stone. The more common practice was to take into partnership one who was acquainted with lithographic procedures. As noted above, during the decade of the 1820s lithography made greater progress in France than in England. It was from the former country, therefore, that American printers received their first introduction to lithography. The earliest lithographic shop in the United States was established in New York City, in late 1821 or early 1822, by William A. Barnet and Isaac Doolittle. The former was the son of Isaac Cox Barnet, who was U. S. consul in Paris from 1816 'til 1833. William studied in Paris, where he apparently met Doolittle. The latter may have been related to the New Haven copper engraver, Amos Doolittle, but there is no confirmation of this. Barnet and Doolittle apparently became interested in lithography while in Paris, where they acquired a basic knowledge of the technique. They purchased a supply of stones and lithographic materials, which they took back to New York, and set up shop at 23 Lumber Street.

An article entitled "Notice of the Lithographic Art," in the 1822 issue of the *American Journal of Science and Arts,* excerpted data from Senefelder's *Complete Course of Lithography* and described Barnet and Doolittle's establishment. "All the drawings in the present number," the article states, "are printed on stone by Messrs. Barnet and Doolittle."[66] Several of the illustrations carry the inscription "From the Lithog. Press of Barnet & Doolittle." This credit line does not, however, appear on the one map in the volume, but it unquestionably was also produced by lithography. The page-size map, titled "Barton on the Catskills," illustrates an article "Notice of the Geology of the Catskills; by Mr. D. W. Barton of Virginia—with a Plan" (Fig. 4.7). The plan shows the Catskill Mountains, the towns of Hudson, Catskill, and Kinston (*sic*), and diagrammatically maps fourteen geologic strata. This appears to be the earliest example of American lithographic cartography. Wood believes that the geological sections were also "the first American scientific illustrations in the new technique."[67] No separate maps published by Barnet and Doolittle are known, and the partnership apparently did not endure for more than a year or two.

Among other early New York City lithographers was Anthony Imbert. As a French naval officer he was captured by the British. While imprisoned in England he studied marine painting and, perhaps, lithography. Upon his release Imbert emigrated to the United States. Shortly after his arrival in New York City, probably in 1824, he was commissioned to prepare, on lithographic stone, illustrations for Cadwallader D. Colden's *Memoir,* which commemorated completion of the New York canals.

An appendix to the *Memoir* notes that "a considerable number of the printed plates of this work are in Lithography . . . these impressions are from the very first press, which on this side of the Atlantic, has been put into effectual operation, many abortive attempts having been made prior to Mr. Imbert's successful one."[68] Among the Imbert lithographs are four maps. The appendix continues,

with reference to the Memoir, it was necessary to have at least two Maps; one of the State of New York, exhibiting the course of the Canal, and its relation to the neighbouring country, but especially to the navigable waters of it; the other to show its connexion, not only with the water courses of the United States, but those of the whole Northern Continent. For this purpose, the artist . . . was charged with preparing Drafts of Maps, to suit the size of the Book, and calculated to be executed in Lithography by Mr. Imbert; all of which have been executed, and the result is seen in those respectable specimens of Lithographic mapping; they are, in many respects, superior to the general style of copperplate maps, for utility and effect, and rivalling it in neatness. . . . There is another map [the appendix continues], exhibiting what was the state of knowledge of the geography of our State, about a hundred years since . . . it is a fac-simile of the Frontispiece to Governor Colden's "History of the Five Indian Nations". . . . If the Committee had done nothing more than to have republished this Map they would have deserved well of their country. This Lithographic Map was executed by Mr. Imbert in one of his happiest moods" (fig. 4.8).[69]

Also in the *Memoir* is a lithographic facsimile of a panoramic perspective of New York City, the original of which was drawn around 1635. On the same page is a reproduction of an early map of Rensselaer's Wyck. This facsimile, and the reproduction of Governor Colden's map of the five Indian nations, may be the first map facsimiles prepared by lithography in this country.

Anthony Imbert issued several other maps as well as a number of lithographic prints between 1825 and 1834. His facsimile of the map of Rensselaer's Wyck, noted above, was also published in Joseph Gates's and Joseph W. Houlton's *History of New York State.* In 1828 Imbert published a cartoon map that is noteworthy as being the first American political cartoon printed by lithography as well as one of the first lithographic maps of the United States.

What appears to be an early one-shot lithographic-cartographic effort is H. Ball's *Map of the military bounty lands in the State of Illinois shewing the true boundaries of each county, as fixed by the Legislature in 1825.* It was "Drawn on Stone & Printed by M. Williams 65 Canal St. N. York 1827." Groce and Wallace identify Michael Williams as a lithographer who worked in New York City from 1824 to 1834. No other maps by Williams are known. Peter Maverick (1780–1831), son of copperplate engraver Peter Rushton Maverick (1755–1811), added lithography to his New York City engraving shop around 1824, and published lithographic prints, and perhaps some maps, until 1831. Prosper Desobry was another early New York City lithographer who followed that profession during the period 1824–44. Among his cartographic works is a large (26 x 40 in) *Map of Brooklyn, Kings County, Long Island, from an Entire New Survey by Alex͏ʳ. Martin, 1834.* At the time Desobry's shop was located at the corner of Broadway and Cortlandt Street.

The early years of lithography in America coincided with the period of canal and railroad building. This, of course, generated a demand for inexpensive maps, which was met by the new printing technique. Representative of canal and railroad cartography in this period is a *Map of canals and rail roads between New-York, Philadelphia, Baltimore and Washington, from authentic maps and actual surveys* (1834). It was produced by "P. A. Mesier's Lith. 28 Wall St. New-York." The map, which was printed from three stones, measures 13½ x 70 inches. Peter A. Mesier did lithographic printing in New York City during the early and middle eighteen thirties.

Bauman, Groce and Wallace, and Silver all record that Henry Stone operated a lithographic shop in Washington, D. C., between 1822–23 and 1846.[70] In 1823 Stone prepared thirty lithographic illustrations for James Lovegrove's *Timber merchant guide,* but no maps prepared by him have been identified. Stauffer believed that Henry Stone "was doubtless connected with Mr. and Mrs. W. J. Stone—possibly a son," while Groce and Wallace think it more likely that he was a brother, or cousin, of W. J. Stone. William J. Stone was born in London in 1798 but came to America with his parents six years later. Stone settled in Washington, D. C., in 1815 and resided there until his death in 1865. He seems to have been engaged in both engraving and lithographic operations and the credit "W. J. Stone, Sculp." appears on a number of maps published between 1825 and 1845. Several early maps of Washington also carry the credit "Engᵈ. by Mrs. W. J. Stone." It is difficult to determine with certainty whether the Stone maps were engraved on copper or on stone. During the early years of lithography the designation "engraving" sometimes appeared on prints or maps reproduced from lithographic plates. However, on some W. J. Stone maps published as late as 1845 plate marks are clearly impressed.

Few of the lithographic firms that operated in America before 1835 survived for more than a year or two. A notable exception was the Pendleton shop that flourished in Boston for more than a decade. Silver confirms that the lithographic "art became permanently established in this country with the founding . . . in 1825 of the firm of William S. and John Pendleton."[71] The Pendleton firm is of particular interest to this study because of the relatively large number of maps it produced during its operative years. William S. and John Pendleton, natives of New York City, were sons of a Liverpool-born sea captain who was lost at sea, in 1798, when both boys were less than three years old. William went to Washington, D. C., in 1819 and worked as an engraver for about a year. He and brother John then traveled west and got as far as Pittsburgh. John, who was an artist, shortly returned to Philadelphia to exhibit at Peale's Museum. William remained in Pittsburgh for several years, teaching music for lack of engraving opportunities. He returned to New York City briefly, in 1824, then moved to Boston where he again obtained employment as an engraver.[72] In 1825 William was in partnership with engraver Abel Bowen.

That same year John Doggett & Company of Boston conceived the idea of preparing lithographic reproductions of paintings, by Gilbert Stuart, of the first five presidents of the United States. Doggett commissioned John Pendleton to take the portraits to Paris where drawings on stone were prepared by a French lithographer, Nicolas E. Maurin. Pendleton also studied lithographic procedures while in Paris, and purchased for Doggett a supply of stones and other essential lithographic tools and equipment. Doggett apparently planned to add lithography to his operations, with the assistance of John Pendleton, but, for some reason, these plans did not materialize. It is possible that the Pendletons acquired Doggett's press and material. John T. Morse, however, reports that, about 1825, William Pendleton purchased a lithographic press from a man named Thaxter, who had purchased it to run off circulars, but the latter did not know how to use it.[73]

A notice, signed by W. S. Pendleton and Abel Bowen, in the 4 and 11 February 1826 issues of *Bowen's Boston Newsletter and City Record*, announced the dissolution of partnership of the firm of Pendleton and Bowen. The newsletter stated further that W. S. Pendleton would continue in business with his brother John, "who will add to the Establishment the advantage of Lithographic Printing. . . . To those whose occasions require Fac Similes, Maps, Circulars, &c., to which this art is peculiarly adapted, Specimens will be exhibited." The effective date of the partnership dissolution was 31 January 1826. However, the December 1825 issue of the *Boston Monthly Magazine* carried a notice on the establishment, by William S. Pendleton, of the first lithographic business in Boston. The account notes that

Perhaps a little before [John Pendleton's] return from France, a few [lithographic] attempts had been made in the city of New York, but they had not reached us, nor have they yet. Mr. Pendleton [the note continues] is a young gentleman of taste and talents, from the State of New York, who was on a visit to Paris, on business of an entirely different nature, and becoming pleased with lithography, put himself immediately under the first artists of France.[74]

The February, March, May, and June 1826 issues of the *Boston Monthly Magazine* include portraits, all of which carry the credit "Lith. of Pendleton." The portraits may have been prepared by a lithographer named Dubois, whom the Pendletons brought to America from France. Peters states that Dubois was the first lithographic pressman in America, although Anthony Imbert may have a stronger claim for primacy.[75]

What may be the first cartographic effort of the Pendletons is a facsimile reproduction of the "Wine Hills" version of John Foster's woodcut map of New England, which was originally published in 1677. The Pendleton lithograph, based on a redraft by M. Swett, is the frontispiece in Judge John Davis's annotated edition of Nathaniel Morton's *New England's Memorial* (1826). As might be expected in this prephotography period, the Swett drawing includes a number of minor variations from the 1677 original.

In 1829 a committee was appointed by the Massachusetts legislature to consider the question of "procuring such a map or such maps of the Commonwealth as the public good requires."[76] The resulting act, passed in March 1830, required towns and cities to have minute and accurate surveys of their respective territories made within a year. Basic triangulation and compilation of the state map were under the direction of Simeon Borden. Fitting together the uncoordinated city and town surveys proved to be an impossible task. Borden reported.

after I had completed the field work, and had calculated a sufficient number of the main triangles . . . I commenced the work of compiling the map, when I found the town maps which had been returned to the Secretary [of State] so incorrectly drawn as to render it impossible, in their actual state, to make a satisfactory map from them.[77]

Borden found it necessary to go into the field, with the result that the surveys were not completed until 1839. Compilation and engraving required five years, and the map was not published until 1844.

Although the town surveys lacked the essential controls and accuracy for use in compiling the state map, the information they presented had considerable local interest. Accordingly, between 1829 and 1836 Pendleton's Lithography printed more than twenty-five of the Massachusetts town plans. Some were hand-colored, and many carried names of landowners. They thus anticipated the town and county land ownership maps that were produced in great numbers after the mid-forties. The Senefelder Lithography Company of Boston also printed several Massachusetts town plans. This lithographic firm was established in 1828 by William B. Annin and George G. Smith as an adjunct to their copper-engraving business. In 1830 the Senefelder Lithography Company was absorbed by the Pendleton firm.

John Pendleton, who was of an artistic temperament, apparently tired of the routine in the Boston shop after a few years and moved to Philadelphia where, in partnership with Francis Kearny and Cephas Childs, he formed the first lithographic firm in that city in 1829. He did not remain long in Philadelphia, however, and by 1830 was back in New York City. William Pendleton, who possessed great business ability, meanwhile continued to operate the Boston company, which became the largest and most successful lithographic establishment in the country. Although maps constituted a major product, the shop also printed portraits, sheet music, fashion plates, book illustrations, and every conceivable kind of work to which lithography was applicable.[78] In 1836 William Pendleton sold the business to Thomas Moore.

A fifteen-year-old boy, Nathaniel Currier, was employed as an apprentice by the Pendletons in 1828. Currier remained in Boston in the employ of William Pendleton until 1833, when he went to Philadelphia to work with M. E. D. Brown, a master lithographer. The following year Nathaniel moved to New York City where he and John Pendleton planned to establish a partnership. Before this was consummated, Pendleton received another offer and sold his interest to Currier. The latter then entered into a short-lived partnership with a Mr. Stodart. In 1835, at the age of twenty-two, Nathaniel Currier established his own shop at Number One Wall Street.[79] Peters writes that Currier's earliest known print, *Ruins of Planters Hotel New Orleans,* was published in 1835. The same year he issued a *Map of Cleveland and its Environs. Surveyed and Published by Ahaz Merchant. October, 1835* (fig. 4.9). Probably because of his early training in map lithography under the Pendletons, Currier printed a half-dozen or more maps during the next several years. They include *Map of the Western Land District Wisconsin, 1836, Map of the northern part of Illinois and the surveyed part of Wisconsin Territory* [1836], *The surveyed part of Wisconsin* [ca. 1836] and *City of Detroit, Michigan from late and accurate surveys, May 1837.* Bland lists three Currier maps (none of the above), among them a *Map of Property at Jamaica L.I. belonging to Abrm. H. Van Wyck.* This is credited to "Stodart & Curriers Lithography" and was, therefore, issued in 1834 or 1835.[80]

Currier, in common with other early lithographers, also prepared maps for book illustrations. One impressive group of such maps illustrates James F. Smith's *The Cherokee land lottery, containing a numerical list of the names of the fortunate drawers in said lottery, with an engraved map of each district,* which was published in New York in 1838. Fifty-nine of the sixty-one page-size maps have the credit line "N. Currier's Lith. N.Y."

On 13 January 1840 Currier published his historic print of the fire on the steamboat *Lexington.* This dramatic illustration caught the public fancy and thousands of copies were sold. It established Currier's reputation as a printmaker, which endured for almost half a century. While he henceforth focused primarily on illustrative and newsworthy prints, Currier did not fully abandon his interest in maps. Among particularly desired items today are the impressive panoramic maps of some eight or ten American cities that were published by the firm of Currier and Ives in the 1880s and 1890s.

It is not possible to note all the American lithographers who produced maps between 1830 and 1850. Among those who deserve more than passing mention, however, is Peter Duval. A native of France, Duval came to the United States around 1828, settling originally in New York City. Some authorities place him in partnership there with George Lehman from that date until 1831, when Duval was employed by the Philadelphia firm of Cephas G. Childs and Henry Inman. Lehman apparently also joined the company and, in 1834, he and Duval took over control of Childs and Inman. From 1839 to 1843 Duval was in partnership with William M. Huddy.

Duval early carried on experiments with color lithography. In 1843 he and

J. H. Richards produced a successful colored lithograph that was published in the April 1843 issue of *Miss Leslie's Magazine.* He was also a pioneer in zincography and, in 1849, was the first printer to install a rotary steam press in his plant. Early examples of Duval's lithographic cartography are *Plan of the City of St. Louis,* by René Paul, printed in Philadelphia by Lehman and Duval in 1835, and *A map of the Georgia Railroad and the several lines of railroad connecting with it,* published in 1839. During the next forty years a large number of maps were printed on Duval presses. The firm was heavily involved in printing county land ownership maps and city plans in the pre- and post-Civil War years.

Although France was America's major source for lithographic information and trained technicians, Germany also contributed to the development and growth of the art in the United States. Of particular interest to cartographic lithography is the firm of A. Hoen & Company, of Baltimore, established in 1835, which is still an active map producer. The company was founded by Edward Weber, a native of Germany, who had learned lithography from Alois Senefelder. His nephew, August Hoen, joined the firm early in its history and, after the death of the founder in 1848, Hoen assumed leadership. Five years later the name was changed to A. Hoen & Company. Also in 1848, the company, on contract with the United States government, printed its first map, which was prepared by Brigadier-General Stephen W. Kearny to accompany W. H. Emory's *Notes of a military reconnaissance from Fort Leavenworth . . . to San Diego.* During the succeeding century and a quarter A. Hoen & Company have printed hundreds of maps, many in multimillion editions.

Official mapping agencies were still in their infancy when lithography was introduced in the United States. Nonetheless, the new medium was early employed to print government maps. Among the earliest examples are certain of the maps, charts, and plans prepared to accompany reports of the two houses of Congress. Because the printing was contracted to various private firms, some maps were reproduced from engraved plates and some from lithographic stones. In the late 1820s and early 1830s Pendleton's Lithography printed several maps. P. Haas, a Washington, D. C., lithographer, also prepared a number of these official maps. Most of the maps published for Congress in the late thirties and early forties, however, carry the credit line "W. J. Stone Sc. Washington City." Some of the maps, at least, appear to be lithographs. As noted above, Stone was both an engraver and a lithographer. An excellent index to early Federal maps is *Descriptive catalog of maps published by Congress 1817–1847,* compiled by Martin P. Claussen and Herman R. Friis, and published in 1941. A revised and expanded edition is in press.

In a report submitted to the secretary of the navy in 1831, the navy commissioners expressed the opinion that

a lithographic press would be the cheapest mode, and would ultimately prove a measure of economy, as we should be enabled by it to provide all the charts and multiply to any extent all the plans, drawings, views, or manuscripts required for the service. . . . Such a press [they emphasized] offers so many facilities and would be attended with such advantages that the Commissioners will, with your concurrence, take the necessary measures to procure one.[81]

Not until May 1835, however, was the commissioners' recommendation implemented. "The introduction of this press was the initial step towards chart production at the [navy] Depot, and in the following autumn the first lithographed charts made their appearance."[82] However, between 1837 and 1842 the navy produced eighty-seven engraved charts based on surveys by the Wilkes Exploring Expedition.

Notwithstanding the early use of lithography for printing official maps, major federal mapping agencies tenaciously employed engraving for reproducing most maps and charts up to the period of World War II. In latter years, however, the image was usually transferred from copperplates to stone or zinc for the printing process.

Prior to about 1850, engraving also offered strong competition to lithography in commercial map production. This was particularly true during the years when lithography was limited to direct reproduction from stone. Printing a landscape or portrait from the artist's drawing on stone posed no problem, for it mattered little if the image were reversed when impressed on paper. A map, however, had to be drawn in reverse on the lithographic stone, just as on an engraved copperplate. Maps also usually required cumbersome stones and large printing presses. The stones were difficult to handle, and there was always the risk that a laboriously prepared stone might be broken, with resultant loss of time and money. The early hand-operated presses also allowed a rate of production not appreciably greater than for engraving. Lithographic cartography, therefore, developed slowly in the United States until the art had advanced to embrace the transfer technique, substitution of metal plates for stone, the rotary steam power press, color printing, and photolithography. These developments, for the most part, postdate the chronological limits of this chapter. Several of them, however, were introduced in the United States, experimentally at least, before 1850.

Of particular significance to map printing was the transfer technique, for the cartographer could prepare his drawing on paper in the usual manner, using an appropriate ink. The right facing image was then transferred from the paper to the stone (later zinc), where it was reversed. After proper preparation of the stone the image was reproduced in multiple copies. Senefelder wrote that transfer "is peculiar to the chemical printing, and I am strongly inclined to believe, that it is the principal and most important part of my discovery. In order to multiply copies of your ideas by printing, it is no longer necessary to write in an inverted sense."[83]

The significance of the transfer process was recognized early in the development of the art in the United States. In the section describing the "Lithographic Department," in Colden's *Memoir,* it was noted that the transfer process may

turn out to be the most useful part of the art, whereby confidential and circular letters of any description may be multiplied with the utmost despatch; as was universally practiced by both parties during the Revolutionary War in Europe, and is still exercised by every one of these Governments, as well as by men of business, thereby saving the expence of copying clerks or transcribers.[84]

Notwithstanding this statement, it does not appear that the transfer technique was widely used in cartographic lithography in America before the mid-forties. There is reason to believe that the process, particularly in its application to lithographic cartography, was introduced to the United States through the medium of anastatic printing.

Among those who listened attentively to Michael Faraday's lecture on the anastatic process, in London, on 25 April 1845, were John Jay Smith, librarian of the

Library Company of Philadelphia, and his son, Robert Pearsall Smith. They subsequently read everything that had been published in England on the process, and John Jay secured the American license and patent rights for anastatic printing. The Smiths also purchased essential supplies and equipment and, early in 1846, an anastatic printing shop, under the supervision of son Robert, was established in Philadelphia. Later in that year the Smiths published, utilizing the anastatic process for reproduction, a facsimile edition of Thomas Holme's 1681 *Map of the Province of Pennsylvania*.[85] This was the only map printed by the Smiths' anastatic press. The apparent success of the publication, however, encouraged Robert P. Smith to publish other maps, employing the facilities of such established Philadelphia lithographers as Peter Duval and, after 1856, Frederick Bourquin. For almost two decades Robert Pearsall Smith played a leading role in publishing city, town, and county maps. He supplied essential transfer inks to local surveyors to use in drafting maps. Smith contracted with a lithographic shop to print a number of copies, after the original drawing was transferred to stone or zinc plates. In exchange for a specified number of copies supplied to the surveyor, Smith obtained copyright and other privileges. How successful he was in his cartographic activities is evident from some four hundred or five hundred maps, with which he was associated (about a third of all such county and town maps produced in the period) published between 1847 and 1864.[86]

Fortuitously, lithographic cartography achieved its full potential in the United States when the demand for special types of maps was greatly expanding, in response to various cultural, technological, and historical factors. Between 1810 and 1850, for example, the population of the country tripled. During the first half of the nineteenth century public land surveys were extended through much of the Middle West, westward movement and settlement were greatly accelerated, new states were established, hundreds of miles of roads, canals, and railroads were constructed, and there was an influx of hundreds of thousands of European immigrants. All these activities and events stimulated the use of maps, which accordingly rolled off the lithographic presses in ever-greater numbers. Because they could be produced in quantity, and at low cost, maps became readily available to all potential users. In the United States, as in Europe, lithographic mapping also stimulated geographic and cartographic study and research. By 1850, however, cartographic lithography was still only on the threshold of its ultimate development. The marriage of photography to lithography during the latter half of the nineteenth century opened new and more promising opportunities for map printing and production.

5 | Miscellaneous map printing processes in the nineteenth century

Elizabeth M. Harris

For printing as well as other trades the nineteenth century was a time of great inventiveness. There were many proposals for new ways of printing and engraving—a bewildering number. My task is to find among these processes some order related to map printing. I propose to describe processes grouped by various associations, and to suggest some of the special demands and opportunities that cartography gave to the inventive engraver or printer. The order is only broadly chronological. But in the nineteenth century, of all centuries, chronology is of the greatest importance and adds much of the interest. The man with a mind to improving printing in 1810 thought in terms of wooden presses barely changed for three-hundred years, hand-cut blocks manually inked from leather-covered pads, and a maximum rate of about two-hundred impressions an hour. Fifty years later the inventor had at his disposal photography, electrotyping, fast steam presses with automatic inking, and printing rates of some thousands an hour.

There is little question that most nineteenth-century maps were made by lithography and copper engraving, and that these processes were quite adequate for normal purposes. The "miscellaneous" processes of this paper do not figure large on a quantitative basis, and since ordinary mapmaking was served well enough by the ordinary processes one would expect to find the odd processes used for odder purposes: maps with unusual requirements such as hastily made charts for daily papers, astronomical maps, or military maps. One also finds maps used by the inventors of new processes as a way of showing off some special quality of the process like its clarity or precision. This is always interesting for the cartographer as well as the printing historian, but it is worth bearing in mind some distinction between the occasional processes used experimentally and those that found a niche, however small.

Not all of the available printing processes lent themselves to the particular needs of maps. The fancier processes were less useful than the plain ones. There was little use, for example, for the attractive impressionistic marks of the process called "electrotint,"[1] or for any of the other ways of reproducing brush or soft chalky marks. One would not expect to find many maps made by aquatint or mezzotint. On the other hand, the promoter of a mid-nineteenth-century process called "acrography" tried to talk his way around the fact that acrography was a plain, quite characterless medium. The artist who dared to use acrography, he said, must be good: "Acrography does not flatter the artist."[2] But for mapmaking such a process might be ideal.

Around 1820 there was a good deal of correspondence in journals about the merits of steel engraving. The hardness of steel offered a very long run of prints—much longer than copper. This was most important for the printers of bank notes needing to issue perhaps fifty thousand identical notes. But steel was too hard to be engraved readily by hand. Using the process called "siderography,"[3] engravers could work on small plates of soft steel, harden the steel, and then transfer the design by pressure to a soft steel roller, harden that in turn, and use it as a master from which to transfer the design to any number of soft steel plates for printing. With larger plates this transfer process was not possible because the risk of damaging the plate in the hardening process was too great. For bank note printing, and later for postage stamps and other forms of "security printing," steel engraving was here to stay. For half the century it also had a place in book illustration and in popular prints issued in large editions. In mapmaking it remained somewhat exceptional, probably finding most use with popular atlases that could expect long runs and a number of editions.[4] However, steel engravings and copper engravings usually look identical, and the only way of telling them apart is by the inscription "engraved on steel" or "engraved on copper." The vast majority of maps have no such inscription, so the extent of the use of steel must remain debatable.

PUNCHES AND MOVABLE TYPE

Mapmaking, like music printing, demands a degree of uniformity and repetition of symbols. Obviously the code of marks must be consistent all over a map. This suggests the use of special tools or mechanical aids, instead of freehand drawings. For example, uniform marks can be made on an intaglio plate with punches. For relief printing, some or all of the map can be set up in movable type. Both methods were used in music printing, and both were attempted, with less success, for maps.

Punching on pewter was a common way of printing music in the eighteenth century. Maps would seem to lend themselves to the same technique, and yet it was unusual for anything more than the dots and smallest circles to be punched. One might guess that pewter was too soft to withstand the editions needed by the map publisher, and that copper was too hard for the larger symbols, such as trees or churches, to be easily punched. But Diderot's *Encyclopédie* of 1787 raised quite different objections. Diderot illustrated and described punches bearing elaborate building symbols, but he condemned the technique because the punched marks were heavy and printed badly; punching gave no tonal variety so trees, for example, came all of the same tone and form and consequently "ne jouent pas assez"; and punching stretched the copper and distorted the map (fig. 3.5). Diderot's[5] objections may tell us more about contemporary taste than about the punching technique. Engraved maps were, he implied, to be judged by the same criteria as any other engraving, as works of art. Mechanical regularity was inartistic, and therefore undesirable.

An exceptional use of punching was that of Francis P. Becker of England who, in the middle of the nineteenth century, introduced an engraving technique specifically for maps that he called "omnigraphy." All separate symbols—hills, towns, lettering, buildings, and even the units of hill shading—were produced by punches, and only the continuous lines were engraved. The results were very clean and clear. The process was known as "Becker's Patent Omnigraphy," but I have found no evidence that it was patented. It was said that the British Ordnance maps were made by his process.[6]

The counterpart to punching in intaglio was white-line engraving in relief—that is, cutting or punching the roads, symbols, letters, and everything else into the surface of a wood or metal block so that in printing they appear white within a black or colored ground. This method was not new in the nineteenth century, but, like punching on copper, it was less useful in practice than one might expect and its application was limited. The method seemed appropriate for certain purposes: astronomical maps, in which white stars against a black sky look right and proper, and town maps with white roads among black city blocks. Among star maps a tradition of white-line engraving went back to the seventeenth century, but the early maps were often rather crude. There is a limit to the amount of information that can be given in white line on black, because a detailed white line image is apt to be more confusing and less legible than the same in black line. There are also technical difficulties for the printer. Any letterpress printer knows that it is not easy to get a satisfactory print from a wide unbroken relief surface. Any unevenness in inking or irregularity in the surface of the plate or the press-platen shows up. The ink itself must be excellent, and the pressure must be heavier than for an ordinary black-line print because there is more surface to be pressed.

The composition of maps from movable type was another old idea that was practiced occasionally and in various degrees of technical purity throughout the nineteenth century. As a simple country cousin of the composed map one must mention the practice of inserting printer's type into slots cut in woodblocks: a trick that was known from the first century of printing and was by no means limited to mapmaking. Another simple form of the technique available to the ordinary printer was that of bending type-high brass rules to make the lines of the map, and inserting letters and symbols of ordinary type. In the late nineteenth century maps of this sort were used for such purposes as realtors posters showing the location of property.[7] But, though the materials were in every printer's shop, the bending, fitting and mitering of rules required a great deal of skill to produce even the simplest result.

Maps composed entirely from movable type found a peak of development in the late eighteenth century in the work of J. G. I. Breitkopf and Wilhelm Haas, and though the technique was used in the nineteenth century the achievement of Breitkopf and Haas was never surpassed (fig. 5.1).[8] The technique called for the cooperation of a printer and typefounder, for the map compositor would need a stock of specially cast design elements and the typefounder's skills were, by the eighteenth century, quite beyond the ordinary printer. From 1773 A. G. Preuschen, a Karlsruhe publisher, made some simple experiments with map composition. Wilhelm Haas, father and son, enterprising Basel typefounders, took his ideas further and published a number of maps, while Breitkopf, a printer-typefounder who was also celebrated for his music type, set up a rival operation in Leipzig.

In 1777 Breitkopf, in an attack on Haas's work, wrote about the difference, in printer's language, between *composing* and *fudging* typographic maps.[9] The fudged work was (and still is) that in which the types did not fit together as tightly as they should and the compositor had to pack the pockets around them with slips of metal and paper or shave down the sides of the types in order to lock them in their places. Breitkopf's comments were directed against his rival, but they underline the immense technical problems that faced any map composer. The printer setting up conventional text has the fairly simple task of spacing characters and justifying lines in a horizontal direction: the vertical arrangement takes care of itself. In a map, however, lines may cross, break, bend, or continue in any direction, and the compositor must have great flexibility from his type and spacing material if the map is not to be severely schematized. Preuschen estimated that a map would require about three-hundred sorts of type.[10] An ordinary font of upper- and lower-case figures and punctuation at that time contained about one-hundred sorts.

Wilhelm Haas the son continued his father's work only until 1803 when the last of his "typometric" maps was published.[11] Other Europeans made occasional use of the idea: McMurtrie mentions Periaux (1806), Douillier (1840), Duverger, and later Jaymes, all of France; Wegener, Bauerkeller (1832), and Mahlau (1851) in Germany; and Raffelsperger in Vienna (1839).

In 1823 Firmin Didot of the French printing house patented a multicolor map printing process that included composition among other features.[12] I am not aware that the process found any commercial use, but it was most interesting because it was designed particularly for maps, an attempt to provide a cartographic scheme and a printing process that were made for each other (fig. 5.2).

Didot produced a map of France to illustrate the patent. The map was printed in eight impressions of which six were from color plates and two from forms of composed type. In the Smithsonian's copy the yellow block was inked with two tones of yellow, pale for land outside France and darker for lines and for the borders of foreign lands. Didot did not specify how the color was to be printed, but he mentioned that it could be printed from intaglio plates as well as relief. For his sample he apparently used engraved wood or metal blocks, with brass rules for the latitudes and longitudes. The black—which comprised all the lettering and town symbols—was printed from two plates, each entirely composed of separate types. According to the patent specification, Didot used letters cast with the faces lining in three directions—sloping up, sloping down, and horizontal. The rectangular body of type allowed him to turn characters through 90 degrees, and by turning the horizontal letters up or down and using the others in either of two positions, he increased the possible number of angles to seven (none of the letters are standing on their heads). It is possible to account for all the angled letters on Didot's map in this way and to accept that the *body* of every piece of type was in a strictly horizontal line. However, I cannot believe that any *series* of angled letters was produced without using thin spaces above and below each character, interrupting the unseen horizontal lines of lead from which, according to Didot, the printing form was constructed (fig. 5.3).

Didot explained that using two forms for the letters allowed words and symbols to lie close together, or even to touch or cross. This suggests one of the great advantages that lithography—then still quite new—and copper engraving had over the relief methods. The lithographer or engraver could easily inscribe one letter across or within another. The printer working with type in a mortised wood block could not do so without using two forms, for each letter lay within a small inviolable space —the rectangular area of the type-body.

An elaborate color system was part of Didot's claim for patent rights. It was an interesting scheme. He suggested that instead of representing classes of cartographic

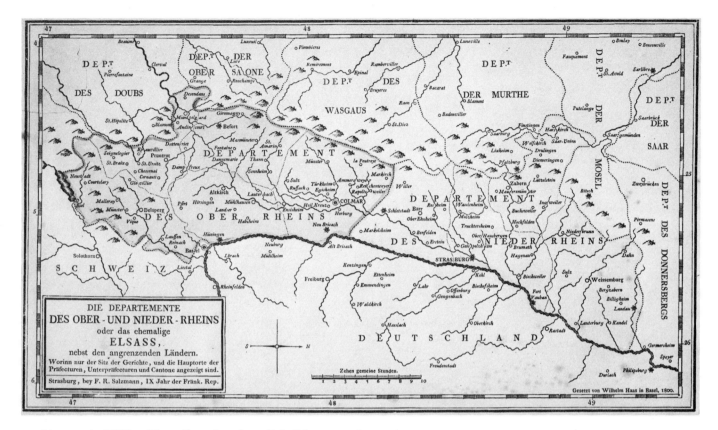

5.1 Typometrie. William Haas, *Alsace,* Strassburg, F. R. Saltzmann, 1800.
Courtesy of the Newberry Library

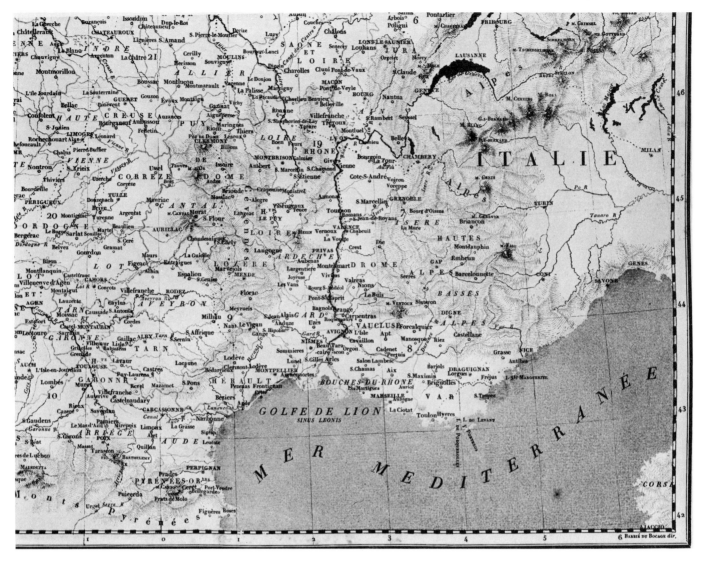

5.2 Firmin Didot, *Cartes Typo-géographiques . . . France,* Paris, ca. 1823.
Courtesy of the Newberry Library

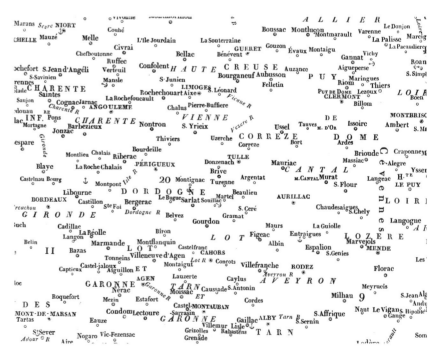

5.3 Detail of type separation used to print fig. 5.2.
Courtesy of the Newberry Library

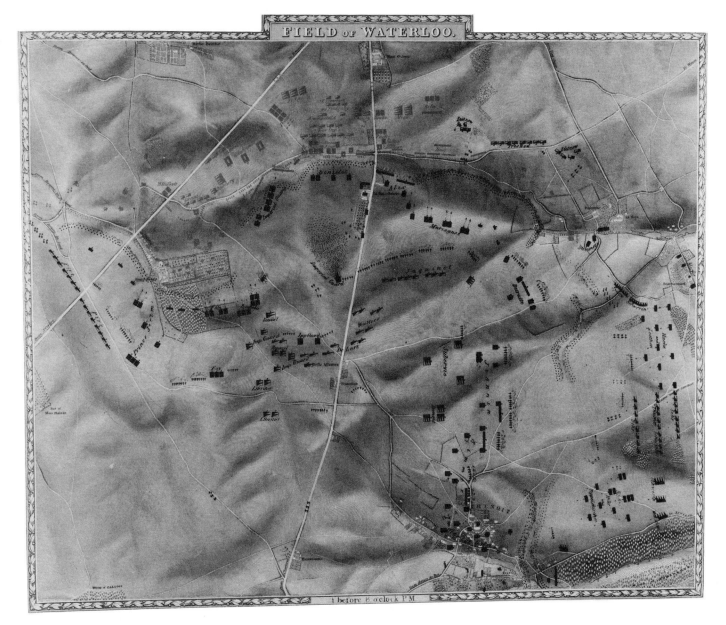

5.4 Medal engraving. "The Field of Waterloo."
Courtesy of the John Johnson Collection, Bodleian Library

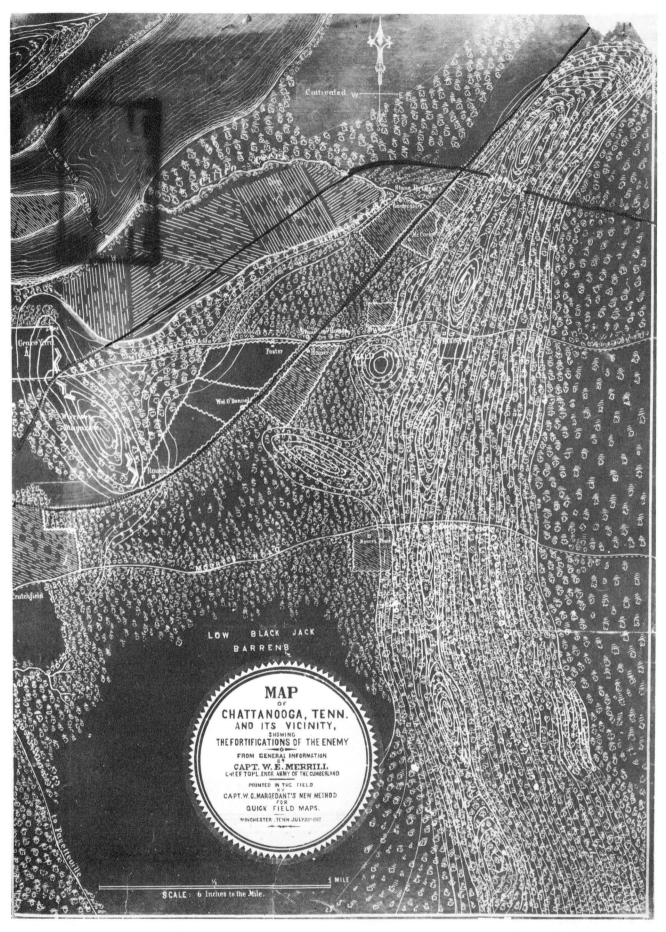

5.5 Margedant's process. W. E. Merrill, *Map of Chattanooga, Tenn. . . .* Winchester, Tenn., 1863
Courtesy of the National Archives

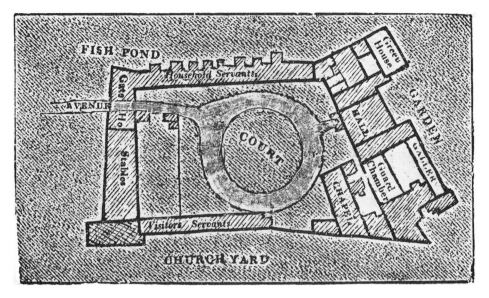

5.6 Nuttall's process, in *Gentleman's Magazine,* London, new series 2 (1834):250.
Courtesy of the Smithsonian Institution

5.7 Cerography. Sidney E. Morse, "Connecticut," in *New York Observer,* 29 June, 1839.
Courtesy of the Newberry Library

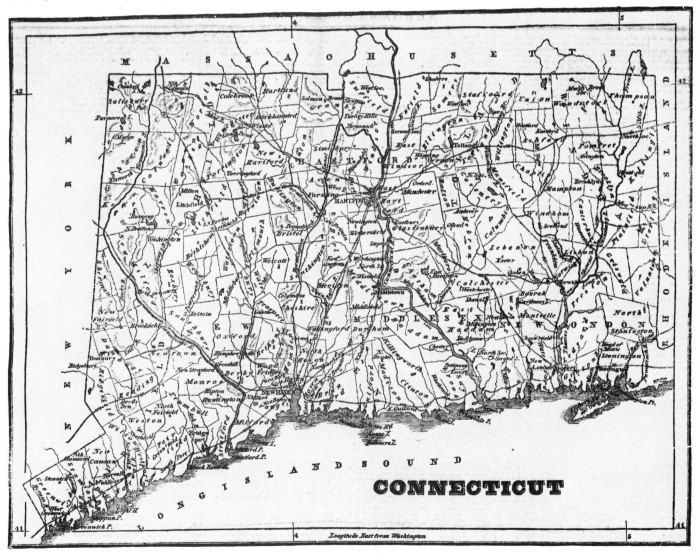

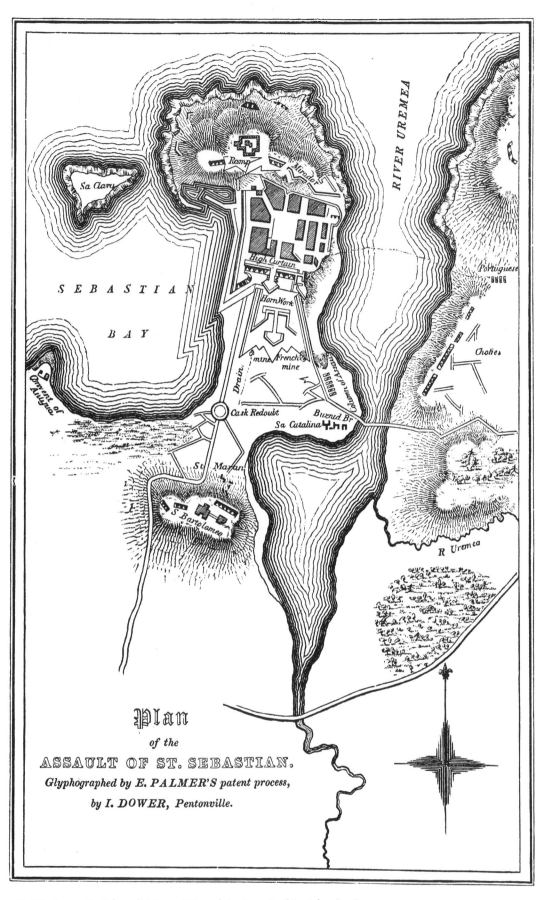

5.8 Glyphography. Edward Palmer, "Plan of the Assault of St. Sebastian,"
in Edward Palmer, *Glyphography; or, engraved drawing . . .*, 3d ed., London, E. Palmer, 1844 (?)

5.9 Wax engraving process.
Courtesy of Rand McNally and Co.

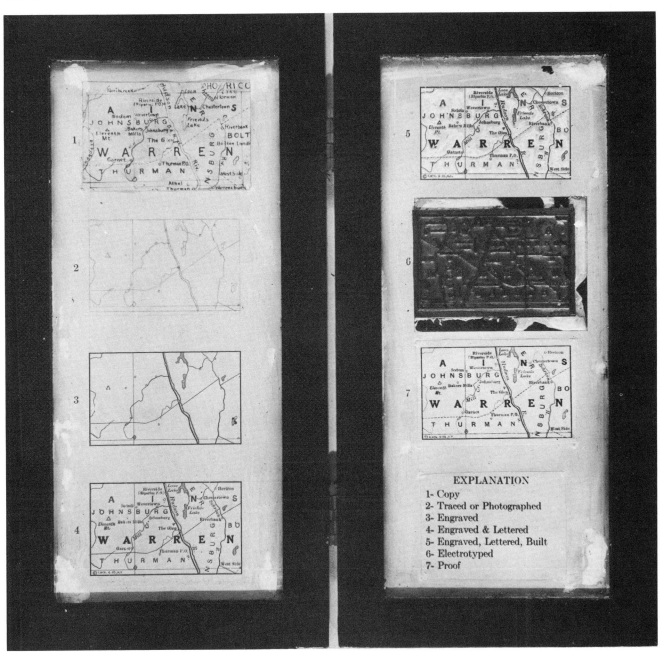

EXPLANATION
1- Copy
2- Traced or Photographed
3- Engraved
4- Engraved & Lettered
5- Engraved, Lettered, Built
6- Electrotyped
7- Proof

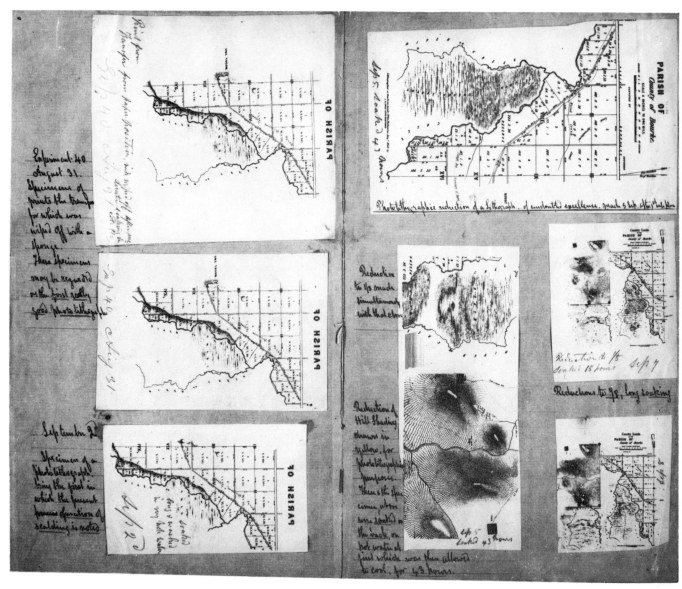

5.10 John Osborne, from a scrapbook of photomechanical experiments.
Courtesy of the Smithsonian Institution

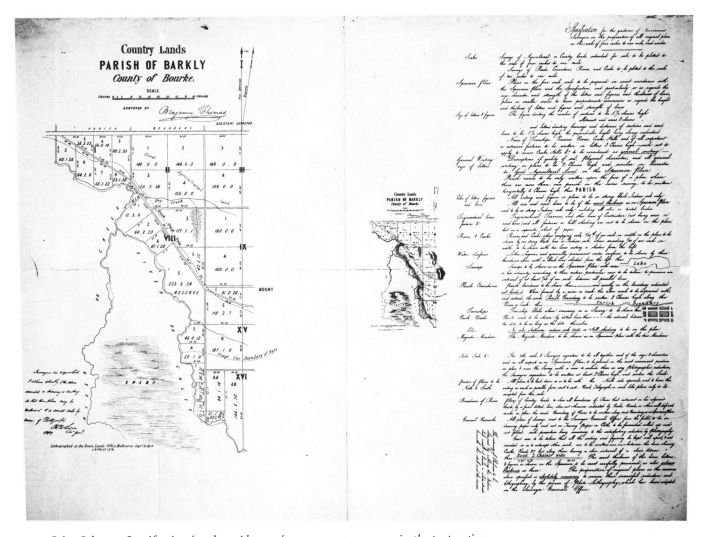

5.11 John Osborne, *Specification for the guidance of government surveyors in the preparation of all original plans on the scale of four inches to one mile, and under.*
Courtesy of the Smithsonian Institution

5.12 John Osborne, specimen of hill shading.
Courtesy of the Smithsonian Institution

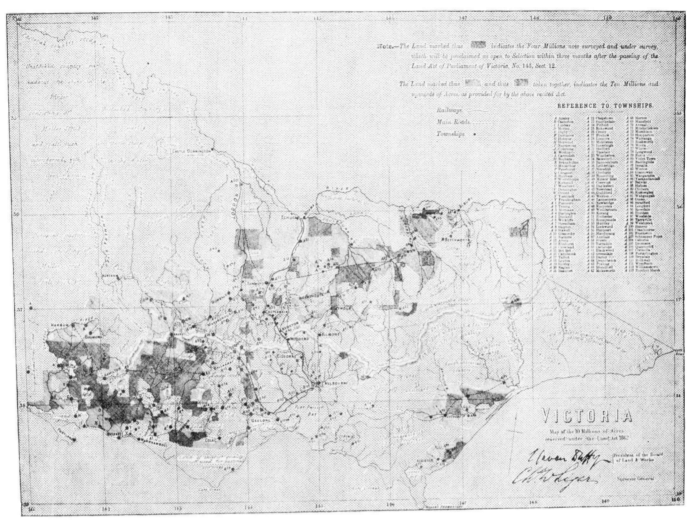

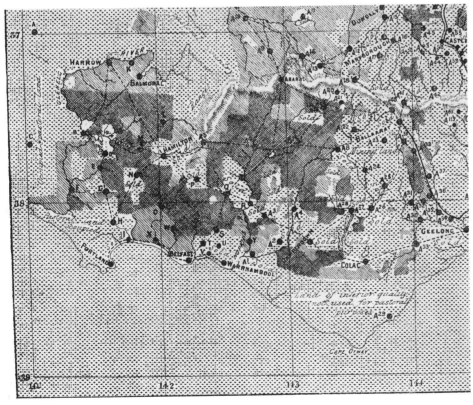

5.13a–b Litho-stereotype. *Victoria.*
Map of the 10 millions of acres re-
served under the Land Act 1862.
Courtesy of the Smithsonian Institution. [Detail inset]

5.14 *Specimen of Photo-Lithography by L. H. Bradford & Co., Boston*, Boston, 1858.
Courtesy of the Smithsonian Institution

features, colors should be keyed to levels of significance and shared by all features. The hierarchy of colors was, in descending order, blue, red, brown, dark green, light green, yellow. Thus the biggest rivers and tops of the highest mountains were blue. Second-class rivers were red, third were brown, and so on. A high mountain showed all five colors on its slopes.

Since exact registration between the colors was critical, Didot suggested an improvement on the conventional press-points. The points were pins fixed in the tympan of the press that pierced the edges of each sheet of paper at its first impression, and were matched to the same pinholes for succeeding impressions, allowing registration close enough for most purposes but not for Didot's. He proposed the use of four pins in a cluster at each side of the sheet, instead of one, to hold it more firmly.

The success of Didot's system depended on excellent block making and printing. The colors of his map are precisely registered, none of the names are entangled, and, as Didot remarked, each town name is centered over its own symbol. But the credit for this should go to superb organization and workmanship rather than to the process itself. In fact, the process must have been less useful just because it demanded such virtuosity.

In 1883 a much more utilitarian version of the composed map appeared in the Boston Type Foundry's type specimen book.[13] The foundry offered "map types," simple arcs, lines, and angles cast on six or eleven point square bodies (about one twelfth and one sixth of an inch high). There were twenty-one characters. Most of them could be turned to give, by my calculations, about seventy-two variations altogether. The types were not designed to be elegant: the description tells us that they were "especially valuable for accounts of fires, accidents, &c. where time is wanting for an engraved illustration." But they did not appear in the following year's type specimen book, despite the note that they were in 1883 already "used in the leading daily papers."

If there is a common quality of the various nineteenth-century processes, great or small, that are covered by this paper, it might be the bias toward mechanization. The old methods of mapmaking involved drawing or engraving by hand; the new should be mechanical. Medal-engraving was a process that illustrates the attraction as well as the pitfalls of this nineteenth-century prejudice.

MEDAL-ENGRAVING

Medal-engraving (or "anaglyptography") was a method of tracing the contours of a relief surface, such as a medal, on to a flat surface by means of an elaborate ruling machine.[14] Medal-engraving machines were developed in France, England, and the United States in the 1820s and began to come into commercial use in the 1830s. The technique was chiefly used for illustrating medals or bas-reliefs, for providing stock ornamental borders for advertisements and trade cards, or for decorating bank notes in a way that was almost impossible to copy by hand (it is still used on bank notes). But the machine was an irresistible toy. It was used rather more playfully for tracing the design of lace and damask silk, for making designs to print on fabric, and, inevitably, for maps.

Unfortunately, engravings made by the machine were always somewhat distorted, the degree of distortion being in proportion to the relief of the original, and unavoidable. The image actually consisted of a series of vertical profiles of the relief, laid side by side on the horizontal surface. Thus any point raised a quarter of an inch from the ground of the relief would lie, in the engraving, a quarter of an inch out of line. This did not deter F. P. Becker who used the process in combination with his punched "omnigraphy" in a demonstration piece, one copy of which is in the British Museum Print Room in W. J. Stannard's "Art Exemplar." Another medal-engraved map is in the John Johnson Collection in Oxford (fig. 5.4). Perhaps a more important reason than distortion for the rareness of such maps is the fact that it was necessary to make a well-finished bas-relief model for the machine to copy, and that must have called for a good deal of extra work. Any corrections involving the *contours* of the map could not be put in by hand on the printing plate: the correction had to be made on the relief model and a new plate engraved from it by the machine.

MILITARY MAPS

Among maps for special purposes I have mentioned astronomical maps and maps for newspapers and posters. The field of war provided another set of peculiar demands and conditions for mapmaking. It is interesting that lithography was promoted as a suitable medium for military purposes, and some of the first lithographic offices were set up at war offices for the production of maps. But lithographic stones were large and heavy, and awkward on the battlefield. During the Civil War the Army of the Cumberland had a lithographic office at its headquarters.[15] At Chattanooga, Tennessee, they had a much more mobile printing office using a process devised by one Captain Margedant (fig. 5.5). Margedant drew his map on thin paper, treated a second sheet with silver nitrate, and made a contact negative print of the first on the second. It seems very simple, but it was not, at the time, all that obvious. Before arriving at this process the army printers had tried ordinary lithography, but found it too cumbersome for the front line, and they had tried camera photography, but this depended on good light and the lens produced distortion at the edges of its field.

Margedant's entire printing apparatus consisted of a box containing several India rubber trays stacking into one another, and a supply of chemicals. Some of the Margedant maps now at the National Archives are marked "40 copies made" or "90 copies made." It was a very simple matter to add to or correct the master copy before making a new edition, and it was reported that sometimes several editions were put out in a single day. Several of his maps have notices like this:

Quick map. July 4 1863.
Note: For all parts marked ? information is wanting. All T. Engineers are ordered to send such immediately to these Head Quarters.

Margedant also produced washable maps, lithographed on cloth, for the special use of the cavalry. Some of his contact-print maps show a cloth texture, suggesting that he made them from one of these lithographed cavalry maps or from a map drawn on cloth.

RELIEF PRINTING

By the middle of the nineteenth century it seems to have been broadly accepted that the future of printing was brightest with letterpress. Lithographers and copper engravers continued to proclaim the merits of their own methods, but the majority of improvements suggested concerned relief printing. The letterpress block could be combined with type and printed at the same time, and, besides, the printing operation was quicker and cheaper than any other. The slowest part of relief printing was in making the block—conventionally by woodcut or wood-engraving. This was the part that attracted the inventors.

Before the introduction of photography and electrotype in 1839 most new relief processes were based on stereotype, or casting.[16] During the 1830s there were a few published examples of plans or other kinds of pictures printed from relief stereotype plates, and rather more reports of attempts that had failed. The essence of these processes was this: A flat surface, such as a metal plate, was coated with a more or less soft composition. The design, or the map, was drawn, engraved, or punched into the composition, each stroke penetrating to the flat base plate. Then this master design was cast and the cast counterpart was the printing plate. Alternatively a plaster cast could be made from the composition, a second plaster from the first, and a metal casting from that.

In abstract, stereotyping was simple and attractive. No special skills were needed in drawing the map, and the design was drawn the right way round (because it was reversed in casting) so ordinary printer's type could be used in punching the letters. It should have been quick, cheap, and easy. It was not so, because of technical problems. The composition had to be soft enough to allow punching or engraving, and yet tough enough to hold its form in casting. The composition was usually spread thinly over the plate so that fine lines could be cut in it, but the final height of those lines had to be enough for the inking rollers to clear the wide spaces between. Several people reported failure because of difficulties with the composition. Others claimed some degree of success, but kept their methods a secret.[17]

P. A. Nuttall's plans in the *Gentleman's Magazine* of 1834[18] are the earliest cast engravings of this kind that I have found (fig. 5.6). Nuttall did not divulge his process. It appears that he found himself, as an editor of the journal, needing some illustrations and having to devise a quick method of making them. His first three published examples of the process were not maps. The next two are the simplest kind of plan but would nevertheless have taken some time to engrave on wood because of the amount of lettering in them. Both maps show a cloth texture in the background, and this is significant. In a relief printing block a wide empty area must either be lowered in proportion to its width, so the inking rollers and paper do not touch the bottom of the cavity, or it must be bridged by some neutral filler such as this textured ground. This was a problem that had to be faced by all relief-block makers, and it will recur in other examples.

The year 1839 saw the introduction of photography and electrotyping—two inventions that offered a wealth of new possibilities to the printers. Photography is the more celebrated, but its effect on printing came later. Electrotyping provided the answer that the would-be casters and stereotypers had been seeking, and it was recognized at once. By means of electro-deposition one could make a copper facsimile of the surface of wax, back the copper shell with lead or type metal, and have a relief printing plate. Relief processes using electrotyping quickly appeared in various countries under various names. We can call them all wax engraving, or cerography. "Cerotyping," in the United States, and "glyphography" (Palmer, England, and Ahner, Austria) were used to a considerable extent for map printing. They were essentially linear processes: tonal variants of the same process (electrotint and *Galvanographie*) were unsuitable because of the ambiguity of their marks.

Probably the first of this family was Sidney E. Morse's "cerography." Morse's first published specimen was the map of Connecticut published in the *New York Observer* in 1839 (fig. 5.7). It was the result of five years' experiment, and one imagines that the invention of electrotype must have been very welcome to Mr. Morse. His work can be taken as the basis of cerography in this country.

Glyphography, Edward Palmer's process, was introduced in 1841 as an autographic process for artists (fig. 5.8).[19] Volkmer Ahner's glyphography was identical and was used at the State Printing Office in Vienna. In England, glyphography, the artist's process, faded and then reappeared in about 1851 under new management as an entirely commercial process used largely for billheads, labels, advertising cuts, and maps. Glyphographs were, I think, always small—book-illustration size. Among the managers and proprietors who were in charge of the process after Palmer were Thomas Best Jervis, a geographer and traveler with a life-long interest in map-making, and Henry George Collins, a publisher of maps and geographical books. From 1851 until 1855 W. A. Blackie was publishing in parts his *Imperial Gazetteer,* which contained 120 glyphographed maps in text.[20] In 1851, too, it was said that Chapman and Hall in England were producing glyphographed maps at a penny apiece.[21]

Cerography, as wax engraving is called in this country, was one of the standard methods of making maps from about 1870 until well into this century, used by companies such as Jewitt and Chandler of New York and Rand McNally of Chicago. The process and its history have been studied thoroughly by David Woodward, and the student is referred to his work. I shall limit myself to a brief description of the technique.

A base plate, usually copper, was blackened with silver nitrate or potassium sulphite. It was heated and coated with a thin layer of a wax composition (beeswax and resin whitened with zinc oxide). The surface of the composition was scraped flat, and an image was transferred to its surface by tracing, drawing, or photography. The engraver then worked on the plate using a variety of gouges, burins, needles, and multiple gravers. Printer's type could be pressed through the warmed wax, and special type holders were devised so that the type should be aligned correctly.

When the engraving was finished the wax on the plate had to be built up so that the final plate would have sufficient depth for printing. This was the alternative to filling broad spaces with a texture, as Nuttall had done. More wax was applied between engraved marks with special building tools: a pen and a hot iron.

When it was built up enough the surface of the wax was dusted with graphite to make its surface conducting, and the plate was put in the electrotyping bath and copper-plated. Only a paper-thin shell of copper was deposited on the wax. The copper shell was removed (the wax master usually being destroyed at this point) and over the back of it was poured a quarter-inch layer of molten type metal. The resulting plate was planed flat and mounted on wood to bring it to type height (fig. 5.9).

PHOTOGRAPHY AND ELECTROTYPE

The U.S. Coastal Survey. During the 1850s the printing department of the Coastal Survey was experimenting with electrotyping and then with photomechanical printing, which they happened to consider the less promising. This department was already committed to copper engraving and intaglio printing, and map work was part and parcel of that commitment. The advantages of new letterpress techniques like wax engraving were not available to them. The superintendent made great efforts to improve intaglio printing to the point where it could compete in time and cost, as well as quality, with letterpress. The results of his experiments were reported year by year.[22]

In 1852 electrotyping was used simply to reproduce engraved plates and occasionally to make new blank plates for the engraver to work on. The following year Superintendent Mathiot reported success in uniting separate portions of engraved plates by electrotyping, thus saving the trouble and expense of engraving anew an entire plate when only a small part of it needed correction.

In 1854 photomechanics entered the scene, but disappointingly. Mathiot wanted to eliminate the "slow and tedious process of engraving on copper" by making "by chemical means, directly from the drawings, engraved plates for printing," and said he was following the instructions of Grove, Talbot, Niepce, Gaudin, and Donné— English and French photographers of the day. He reported two successes out of a hundred. He worked more with Daguerre's processes than with Fox Talbot's. This was unfortunate because the future of photoengraving lay, as it turned out, with processes based on Fox Talbot's work rather than Daguerre's.

In 1856 there was the interesting remark that "the last specimens indicate that a little further improvement will enable the photoelectric process to commence a rivalry with the lithograph, and even to contend with hand-engraving."[23] But this throws more light on the relative status of lithography and copper engraving than on photoengraving, which never achieved that improvement at the Coastal Survey offices.

The reports for the next few years deal chiefly with the success of electrotyping. Plates were joined, extended, cut up, and rearranged. The slowest part of the process was in building electrolytically a copperplate thick enough to be printed under the great pressure of the copperplate press. It was reported (though it sounds unlikely) that a paper-thin copper shell of a map could be backed with a sheet of steel and printed. And in a prophetic comment Mathiot wrote in 1857:

The working of thin electrotypes has suggested to me the idea of using these plates on a circular bed or roller, and gaining thereby the great advantages of cylinder printing for flat plates. This has often been sought before, but the impossibility of getting a rigid plate to conform accurately to a cylindrical figure has hitherto defeated it. As the thin electrotypes are easily strained over a surface, the great desideratum is now attainable. I am about having this matter put to a practical test, and have every hope that copper-plate printing can thus be executed by steam machinery, with almost the rapidity of letterpress work.[24]

This was 1857. High speed rotogravure, which is what he was describing, was not to become a reality for another forty years and one of the chief hurdles was in fact the problem of wrapping an image around a curved surface.

In 1860 photolithography was given some attention. It was dismissed as being "too coarse and mechanical in its nature"—suitable only for second-class maps. This was, of course, the prejudice of a copper-engraving office, but it had reason. In 1860 too, a lithographic office was started at the United States Coastal Survey for the first time, rather reluctantly. Not until 1864, when the demand for maps became very much greater with the war, was it granted that the lithographic office had justified itself.

John Osborne. The Australian Department of Lands and Survey was in much the same position as the United States Coastal Survey, but here lithography was the principal process. Electrotyping had nothing obvious to offer the lithographer, but the chief of the office, John Osborne, began to carry out experiments in photolithography while his colleagues worked on photo-relief processes (fig. 5.10). Osborne designed a photolithographic process specifically for map printing. There was evidently some question about its worth, and a parliamentary commission was appointed to look into the matter. Its findings, again, are most interesting in showing what were the practices in an office such as Osborne's and what were the problems, both human and technical, that would be met by a new process.[25]

Before Osborne introduced his process, Australian surveyors in the field would draw rough sketches that were sent to headquarters and redrawn on lithographic stone or copperplate by professional lithographers or engravers. With photolithography, the surveyors' original maps themselves were reproduced, and they had to be more than mere sketches. The surveyors were required to draw on a larger scale than usual and to use uniform letters (a letter sheet was sent to them to copy, with instructions; fig. 5.11). Back at headquarters a professional draftsman would add only the hill shading and other features whose appearance was considered more a matter of art than of accuracy (fig. 5.12). Then the map was photographed, and here the first technical difficulties might arise. One witness at the enquiry stated "I have known Mr. Osborne to be two or three days waiting for the breaking up of the weather, before he could take a picture . . . it was either too wet, or dusty, or something."[26] Osborne had to work in the open air: under a roof there was not enough light. In spite of problems and opposition from some of the lithographers in the office, Osborne's process was successfully established in Australia and used for the numerous plans for land sales that his department put out. Osborne himself came to the United States soon afterward and started the American Photo-Lithographic Company, a company that specialized, and thrived, in the field of plans for patent specifications, and maps.

Photolithography was not the only way of applying photography to the printing press, and in the fifties and sixties photolithography did not separate itself from other methods as clearly as it does now. By juggling with the photosensitive properties of various gummy solutions one could obtain all sorts of relief, intaglio, and lithographic printing images. Most of the processes adapted themselves more readily to line work than to tonal, and maps provided an excellent way of demonstrating a line process. In the photomechanical collection at the Smithsonian, a collection of pioneer work put together by Osborne, there are quite a number of maps. Nearly all of them are strictly experimental. The examples that follow are chosen from the Osborne Collection to illustrate this period of exploration when the term "photomechanics" was not yet coined and photography was seen as a tool to supplement conventional processes rather than the cornerstone of something entirely new.

Charles Henry, an American colleague of Osborne's, had a relief process based on Osborne's photolithography (1860s). A transfer was made on to metal and etched into relief. The Heliotype Company, most of whose work was with photolithography and collotype, also made a relief map by converting a lithographic form into a relief one. It was not as successful as Henry's.

John Ferres was an Australian who had worked with Osborne. According to a note in pencil on the bottom of his map (post 1862), he etched an image on stone into relief and then stereotyped it. The stone, which is brittle, would be unsuitable for printing at a platen press as is usual for relief blocks. Ferres filled the wide spaces with texture, for exactly the same reason as Nuttall in 1834. Apparently he had difficulty etching the stone to a sufficient depth without this filler (fig. 5.13a and b).

Korn of Berlin and Bradford of the United States both had photolithographic processes that were essentially tonal rather than linear, and therefore did not lend themselves readily to mapmaking. Nevertheless, Bradford reproduced the Foster woodcut map of New England (1864), a map with just the right softness to show off his process. He also used a map of "Photolithographic Bay" as an element in his trade card (1858; fig. 5.14). Korn (1865) used his process most effectively to reproduce the natural shadows on a bas-relief map. The process was expensive because he had first to prepare a bas-relief model, as had the medal engravers. A bas-relief model occurs in another map in the Osborne Collection: the Amakirima Group published in 1873 by the U. S. Navy Hydrographic Office. The map bears this note:

In preparing the topography from the curves and views on file, a plaster model was made and photographed. Maikirima and Assa Ids. are heliographic etchings. The rest are transfers of photographs by hand. A tint was ruled over the whole to aid the highlights.
Model and Engraving by Frh.r. F. W. von Egloffstein.

To the naked eye there is nothing to distinguish the Maikirima and Assa Islands from the rest of the map (which was made, presumably, by an engraver using the transferred photograph as a guide). But under a glass the two islands show traces of a ruled screen in addition to the ruling of the tint that covers the entire map.

Friederich von Egloffstein was the proprietor of the Heliographic Engraving Office and inventor of a photogravure process based on the use of a ruled screen, a forerunner of modern photogravure. On an advertisement dating from about 1865 he stated:

Engravings upon Steel in Line, Stipple, Mezzotint, &c are produced by the Sun from all objects in nature, as Landscape and Architectural views, Portraits, scenes from life and of still life, Microscopic views of Anatomy and Natural History, also reproductions of Manuscripts, Artists Designs, Maps, Diplomas, Labels, Business Cards, Vignettes and Bank Checks, which are made counterfeit proof by the process.

So far as I know, Von Egloffstein did not make other maps. But the heliographic process slipped into some of his other commissioned work such as the vignette of a trade card for the Bay State Iron Company and the background of a group of views of real estate in Putnam, New York.

Other photoengravers used maps to advertise the camera's ability to enlarge or reduce images. Toovey of Belgium used his photolithographic process for a much reduced copy of a copper-engraved map (1863). His process was similar to Bradford's; the image is soft, and much detail is lost. Wenderoth and Burchard of Germany also used photography to change the scale of maps. Wenderoth's maps (1872) is inscribed in pencil *16 mal verkleinert vom Original.* Burchard of the Prussian War Department printed a map in four different sizes, ranging from 2¼ x 3⅜ in. to 9 x 13½ in. (1863). He succeeded in showing that with a sufficiently clear, hard-lined process a map would not suffer from photographic reduction.

In connection with changing scale, and in closing, I should mention one other odd process: the "electro-printing block process" of Henry Collins.[27] Collins was one of the proprietors of Palmer's glyphography in London. In the 1860s he set up his own shop selling maps and advertised his process for enlarging or reducing maps. The formidable name has very little to do with the process. A print was made from the block or stone onto a sheet of rubber stretched on a frame like the tympan of a drum. The rubber was then either stretched further or relaxed, and when the image was brought to the right size it was set off onto stone and printed lithographically.

Collins established a company to exploit his process, the "Electroblock Printing Company," but I have found no maps attributed to it.

Collin's idea seems so simple as to be almost ridiculous, but there was more to it than one might suppose. It takes ingenuity to reduce an image mechanically without the aid of a camera lens. With a pantograph (which was well known in the nineteenth century) the outlines of a map might be traced, but someone would still have to go over them to convert the tracing into a lithograph or engraving. Printing on an elastic surface for transfer was not new: it had been used since the eighteenth century for printing on ceramics. In 1807 a Frenchman, Gonord, used a process almost identical to Collins for changing the size of printed images.[28]

These processes of the "miscellaneous" group were individually delightful. One can all too easily imply that their significance was greater than it was. The importance of a precursor is always questionable. But as a group the processes were clearly symptoms of a certain restlessness of the period. More than that, each one can be taken as a kind of measuring rod telling us something about the state of affairs at that time: the shortcomings of established cartographic printing practice, and what was technically possible without the burden of commercial organization.

6 | The application of photography to map printing and the transition to offset lithography

C. Koeman

The history of photochemical processes has its roots in the eighteenth century.[1] For their practical application in photography, however, we have to turn to the beginning of the nineteenth century, when, in 1814, Joseph Nicéphore Niepce de Saint Victor found that bitumen became insoluble when exposed to light. Already in 1802 T. Wedgewood had found that images could be produced on paper soaked in silver nitrate. Niepce developed his system for practical purposes in partnership with L. J. M. Daguerre, who made picture photography practically applicable. Daguerre used metal plates coated with silver iodide exposed to light with a camera obscura and developed with mercury vapor. The official announcement of the invention was made by D. F. Arago on 7 January 1839 at the Académie des sciences.

Independently, William Henry Fox Talbot investigated the use of silver chloride papers and produced negatives with his camera obscura in 1834. He also produced the first positives by contact printing, the so-called photogenic drawings. Three weeks after the "discourse" by Arago in Paris, Fox Talbot presented his invention to the Royal Society in London.

The year 1839 saw the first use of the word "photography," used in England by Sir John Herschel, son of the famous astronomer, in a lecture at the Royal Society, where he explained his invention of the process of fixing silver images in sodium thiosulphate, which he had discovered twenty years earlier. In 1840 Fox Talbot made another most important discovery: the latent image and the possibility of developing it with gallic acid, in what he called the "Calotype process." Exposure times, which had previously been as long as one hour, were reduced to one to three minutes. Further reductions of exposure time and further improvement of the picture quality were achieved by the introduction of J. Petzval's portrait lens in 1840. But for a period of about twenty years, the daguerreotype remained superior and had considerable commercial success. The nontransparent paper base of Fox Talbot's negative was a considerable barrier to the development of photography. In June 1848 at the Académie des sciences, Niepce announced his invention, producing a light-sensitive coating of albumen on a glass plate. Various improvements followed, such as those by Fox Talbot, but the durability of the plates was short and the exposure time was too long. Finally, in March 1851, Frederick Scott Archer announced the invention of the wet collodion glass plate as a base for light-sensitive silver salts. This was the most important contribution to photoreproduction techniques for about half a century, until bitumen was discovered as a light sensitive emulsion. The wet-collodion plate has remained (be it in exceptional cases) in use until today, but its application diminished with the introduction of stable lithographic film about 1940.

From the middle of the nineteenth century, the history of photography can be divided into two branches: one leading to the perfection of dry-plate photography, the introduction of celluloid as a base for emulsions, and the development of portable cameras and roll films; the other branch leads us to the use of photography for plate-making, a development we will now pursue.

For a proper understanding of the relation between map reproduction and photography we should go back to the beginning of the nineteenth century and try to explore the feasibilities offered to map printing by photochemical reactions, before the invention of the photographic negative in 1839. Contrary to Fox Talbot's photographic experiments, the earlier work by Niepce had been made with reproduction in mind. About 1813, Niepce was looking for a process by which to reproduce by means of the printing press.

Instead of the chalk stone, on which illustrations had to be drawn by hand, Niepce used a tin plate. He had the idea of making the tin plate light-sensitive and exposing the coating to sunlight under an engraving made transparent by wax. After various experiments with sensitive silver salts, he used light-sensitive bitumen dissolved in a solvent such as benzene. After two to four hours' exposure in the sun, the soluble bitumen was dissolved, leaving the light-hardened bitumen as an acid-resisting coating on the plate so that the lines of the picture could be etched into the metal. By this process he obtained an intaglio printing plate that was used for reproduction on a copperplate press. Niepce named his prints "copies de gravure." As far as we know, this was the first application of photography to platemaking for printing. In fact, it was the beginning of photogravure. Unfortunately, maps do not appear to have been printed as "copies de gravure," but this must have been perfectly feasible.[2]

Before the use of reproductive cameras, photography was also used to transfer line drawings onto the light-sensitive coating on a stone or copperplate in order to facilitate the engraving. This contact-printing process made use of various light-sensitive emulsions, among which bitumen has the longest record. The invention of light-sensitive chromium salts dates back to 1839, when Mungo Ponton discovered that paper treated with a bichromate could yield a permanent photographic image. This led directly to the widespread use of bichromated glue or gelatine. Mungo Ponton's discovery was further developed by Fox Talbot, who found that gelatine, when sensitized with potassium bichromate and exposed to the action of light, became insoluble. Soon afterward, albumen, sensitized with bichromate of ammonia, replaced gelatin. This solution was spread over the surface of a stone or zinc plate. To secure an even coating the zinc plate was fixed to a whirler.

1850–1900   About 1850 most of the basic photochemical reactions in use today were known: silver salts, chromium salts, developers, etching acids, etc. However, one vital element was lacking: artificial light. It is amazing to see how slowly the application of artificial light to reproductive photography developed. Even in 1907, technical instructions for building a reproduction plant refer to the "daylight room" as the most important department.

Looking back at the years of the middle of the nineteenth century and remembering the spectacular achievements in cartography, it is hard to believe that the very detailed and meticulously drawn maps at that time were reproduced with very primitive equipment and materials. For example, the mapmaker did not have access to:

Electricity for lighting, exposure, or power;

Transparent drawing films and dimensionally stable drawing materials;

A less expensive printing surface than copper or stone;

Photochemical platemaking processes allowing for long runs (printing down on zinc by the Strecker process was invented in 1901);

Stick-up lettering and symbolization. All the map elements had to be drawn freehand;

Screens for printing tints from one stone;

Rotary presses for running large editions.

From this it becomes clear that maps had to be printed mainly in one color from the copperplate or from stone. In addition, the engraving or fair drawing had to be done at the scale of reproduction. Still, it is true that, notwithstanding the modern equipment of today, the technical virtuosity seen in maps produced between 1850 and 1920 has never been surpassed. This is mainly due to the mapmakers' skill and manual technique, often exercised under military discipline. Engravers and lithographic draftsmen were forced to pack a maximum of detail within a given area. It was the period of the great Stieler atlases, and of overloaded military topographical maps. It was also the period when draftsmen were asked to demonstrate their skill by writing the Lord's Prayer within the circumference of a dime. But it was also due to the invention of various ingenious printing techniques. Most of those techniques were short-lived; they produced some unusual types of maps, very charming in their appearance, as Dr. Harris has shown in the previous chapter.

The first repro-cameras. The first cartographic cameras were installed about 1860, homemade and often built on the spot. Thanks to an eyewitness report of a professional cartographer, we are able to reproduce here what was felt about the problems inherent in map reproduction by means of a camera. This report was written in 1879 by W. F. Versteeg, director of the Topographical Surveys of the Netherlands East Indies.[3] It is one of the oldest technical papers known on this subject. Versteeg deals at great length with the qualities demanded of cartographic cameras:

Distortion-free lenses (he mentions the "rectilinear lens of Dallmeyer, the doublet lens of Ross, the aplanatic lens of Steinhall");

No vibration (the exposure time was ten to thirty minutes);

Parallel planes of copy and negative.

All these requirements are met by modern, commercially available cartographic cameras. The cartographic cameras of the years 1860–1900 were not installed in a darkroom. On the contrary, the camera of the Survey of India worked in the open air, in the blazing, unshaded sun. In Europe and North America, the camera was installed under a shed or in a room with a glass roof (figs. 6.1 and 6.2). According to Versteeg, the first to apply the camera to map reproduction was most probably Captain A. de C. Scott under the direction of Colonel Sir Henry James, director of the Ordnance Survey of Great Britain and Ireland. After the first experiments in 1855, the method was made practicable in 1859.

About the same period, camera photography was brought into use for map reproduction in Australia by John Osborne, in the United States by Charles Henry, and in Belgium, where it was used for the reduction of the military topographical map 1:40.000. Mumford reveals that the earliest application was made in Leipzig, before 1854, published in the book by Adolphe and Hermann Schlagintweit, *Epreuves des cartes géographiques produites par la photographie d' après les reliefs du Mont Rose et de la Zugspitze* (Leipzig, 1854).[4]

One point of particular importance should be stressed: the quality of the map drawing had to be adapted to the camera. This was something completely new to cartographers of 1860—until then, the engraved lines in copper or stone were automatically solid and pure black. It now became the task of the camera to produce line work of similar quality. To that end, the originals had to be drawn on paper and, because the lighting was far from optimal, inked in a rather heavy line.

Electric lighting. It should be understood that Versteeg wrote two years before the Paris Exhibition of 1881, where electric lighting was first used on a wide scale. In 1846, the electric arc light was first used in a production at the Paris Opera. It should be borne in mind that the source of energy at that time was the primary battery. The practical use of electric light had to await the development of the dynamo. About 1860, when the camera was first used for cartographic work, adequate lighting was still not available; for photocopying work, electric light was also an exception. In 1880, Volkmer stated that in England Woodbury used electric light for photocopying in 1866, and in 1884 Eckstein reported that he made successful photographic experiments with electric lighting. But in 1885, when Volkmer added a chapter on this subject to his earlier edition, the only topographical institutes where electricity was in use were the topographical bureaus of the Prussian General Staff, those of the Bavarian General Staff, and those of the Portuguese Geographical Institute.[5] From 1880 onward, the carbon arc light was gradually introduced into the process of map reproduction. It remained in use long after the invention of the high-pressure mercury vapor lamp in 1933. After 1880, carbon arc light lamps were also used for indoor lighting (fig. 6.3). The psychological effect of the new electric lighting was of greater importance than the effect of electric lighting on map reproduction. But in the context of this paper I cannot dwell on this effect. Let me only repeat what William W. Blades remarked in *The Enemies of Books* (London, 1880) when electric light was installed in the reading room of the British Museum. He reported how the new lighting was "a great boon to the readers." As disadvantages he mentioned the "humming fizz which accompanies the action of the electricity" and the falling down of particles of hot carbon on the bald heads of readers, a nuisance remedied by placing collecting pans under each arc light lamp.

Nowadays it is forgotten that the first applications of repro-cameras to map reproduction was the production of photographs of maps and *not* the production of pho-

6.1 Cartographic camera in use at the Military Topographical Surveys of the Netherlands about 1880. Installed outdoors, as usual, to receive maximum daylight, and protected by a shed against the weather.
Courtesy of the Topografische Dienst, Delft

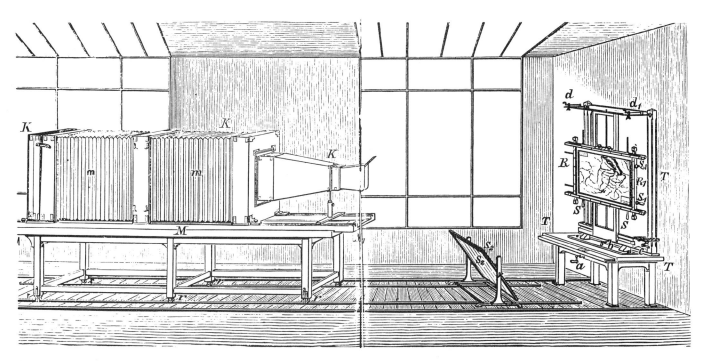

6.2 Cartographic camera, ca. 1880, in O. Volkmer, *Die Technik der Reproduktion von Militär-Karten und Plänen,* Vienna, 1885

6.3 Carbon arc light, ca. 1880, installed in a portrait atelier, in O. Volkmer, *Die Technik der Reproduktion von Militär-Karten und Plänen,* Vienna, 1885

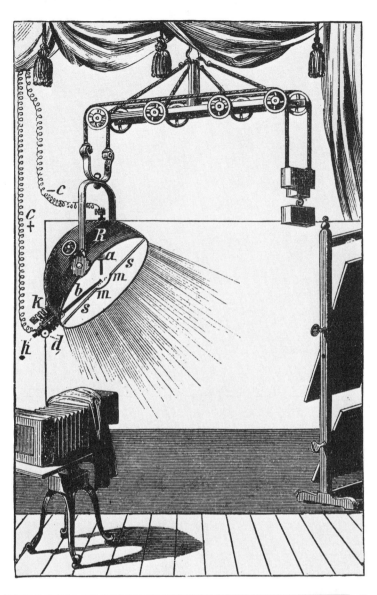

6.4 Line-engraving machine used in the 1876. Eckstein process,
Courtesy of the Topografische Dienst, Delft

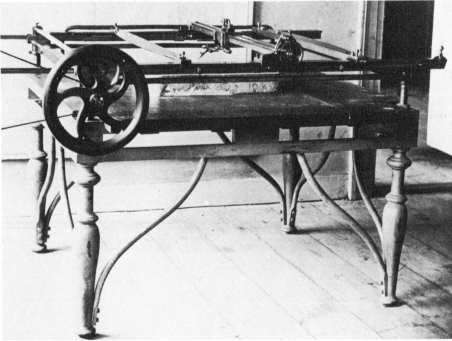

6.5 Color printing by the Eckstein process, from C. Eckstein, *New Method for reproducing maps and drawings,* International Exhibition, Philadelphia, 1876

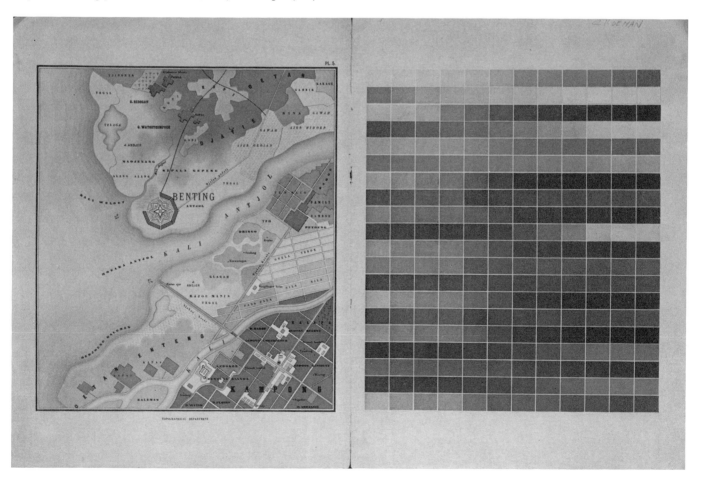

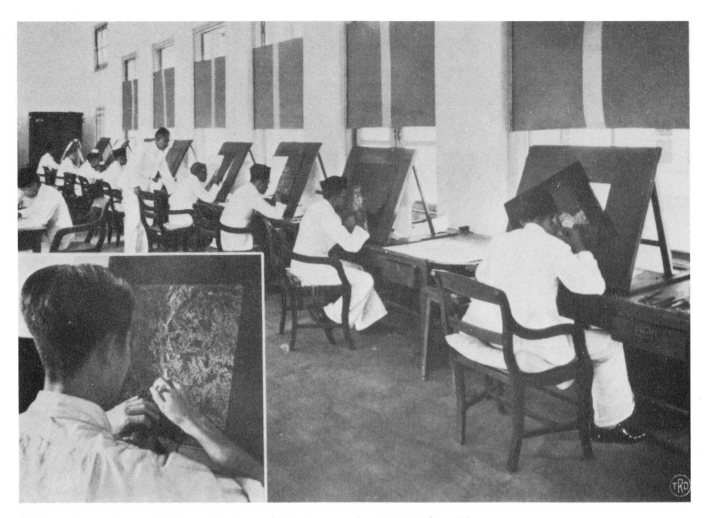

6.6 Engraving on glass at the Topographic Bureau of Batavia, 1927, from Th. J. Ouborg, "Art reproductiebedrijf," in 75 *Jaren Topografie in Nederlandsch-Indië,* Batavia, Topografische Dienst van Ned. Indië, 1939

tographically obtained master plates or stones from which to print them. Photographic duplication was considered to be an alternative to map reproduction, for in 1870, just like in our times, certain types of large-scale maps required only a very limited number of copies, say five or ten. In addition, by using the cartographic camera, it was no longer necessary to draw at the scale of reproduction, and the use of the rather unstable hand pantograph was obviated, generally increasing geometrical accuracy of maps. But many problems of photolithography had to be solved.

The state of map reproduction about 1875. The processes in use were:

Copper engraving, at that time still preferred as the best technique, but suffering from a shortage of qualified engravers to meet the increasing demand for faster and larger printing runs;

Lithography, either on stone or zinc, subdivided into gravure, drawing with lithographic ink, and drawing with crayon;[6]

Anastatic printing.[7]

About 1875 the role of photography in lithographic printing was limited to the following applications:
1. Contact-printing a glass negative on the light-sensitive coating of a stone or zinc plate. The positive copy served as a guide for ink drafting or engraving on the stone or zinc.
2. Transfer (German, *Umdruck;* French, *faux décalque*) of a photographic paper print on the surface of the stone or zinc. The print contains a gelatine emulsion, developed with a greasy lithographic transfer ink. After developing, the image of the print is transferred to the surface of the stone or zinc under a hydraulic press. The stone is then ready for printing. This method was the regular process to transfer a drawing on the surface of the stone. Before the application of photography such a drawing had to be made by hand with greasy ink on paper. It thus becomes clear that the qualification "photo" added to the word "lithography" at that time, about 1860, implied the photographic production of the transfer-document. In later years, about 1903, after the invention by Dr. Strecker, the name photolithography was applied to the chromium-albumen platemaking method by means of actinic light.

Among the solutions of serious problems encountered in the map printing processes around 1860 were:

The substitution of heavy stones by a lightweight printing surface;

The printing of several tints from one stone;

The replacement of the time-consuming manual engraving process by photographic engraving.

Those solutions were only acceptable if they guaranteed the same high standard as copper engraving. Solutions for each of the three problems were found by the military surveys in France, the Netherlands, and Austria. France developed *"zincographie"* to a very high perfection. The Netherlands invented the first systematic three-color lithography. Austria invented copperplate engraving by means of helio-gravure. Let us report on each of these new techniques.

"Zincographie." Mumford, in his article on the map reproduction processes used in England, distinguishes between zincography and photozincography.[8] The original meaning of the term zincography is printing from a zinc surface on which the map has been drawn with lithographic ink or was transferred from an original. The Ordnance Survey of Great Britain applied photozincography for making facsimiles of early maps and documents about 1860.

For a long time in the nineteenth century, the *Carte de France,* scale 1:80,000, on 274 sheets (successor of the famous *Carte de France* by Cassini, 1:86,400), was printed from the original copperplates. When a new map was needed, several experiments were carried out that resulted in a new *Carte de France,* scale 1:50,000, on 950 sheets, printed from zinc plates in five colors, to which a hill-shading plate was added. The use of zinc was clearly important: "On peut dire que c'est l'emploi du zinc comme support d'impression, permettant un tirage rapide à froid et sans déformation, qui décida, en 1879, de l'avenir de la Technique Cartographique en France."[9]

Contrary to what the name suggests, zincography is different from our well-known modern technique of lithographic printing from a zinc plate. For the plate-making process, a light-sensitive coating of albumen with chromium salt was applied to the zinc, which was exposed under a glass negative of the line drawing. After exposure, the plate was inked and afterward washed to remove the non-hardened albumen emulsion. These plates were only used for printing proofs on special paper, which were afterward transferred to the surface of the final zinc printing plate. Obviously, the emulsion was not hard enough or the surface of the zinc repellent enough to water to enable the printing of more than one copy.

The next step in the zincographic process was the preparation of the surface of the final zinc printing plate. By a lamination with phosphoric acid and gallic acid the surface was rendered repellent to printing ink. To this surface, the proofs were transferred under a press. Then followed the final step: the engraving by hand of the drawing in the zinc. But here engraving was not done as in copper engraving— not a deep cut with a burin, but just a light scratch with a steel needle. This light touch with the needle was enough to produce a fine groove to retain the greasy ink. Before inking, the plate was rubbed with oil; the oil did not stick to the repellent surface. Because only the scratched parts accepted ink, no water was required as in our well-known lithographic process. Being a dry process, it guaranteed excellent register of the various colors. Further, the fineness of the lines surpassed lithography from stone.

A variant of the zincographic process consisted of exposure under a diapositive instead of a negative on zinc coated with a solution of light-sensitive bitumen. That part of the coating that was not hardened by the exposure was washed off and subsequently the plate was lightly etched. But it is curious to read that the plate was not used for printing, but just to produce keys to guide the engraving of the black, blue, and red plates.

The Eckstein process.[10] This was a flat color map reproduction process, invented about 1860 by Charles Eckstein, director of the Dutch Topographical Department of the Ministry of Defense, twenty years before the invention of line screens by Meisenbach (1882). The combined prints from the three stones (one for each primary color: red, yellow, and blue) rendered possible the printing of twenty to forty different colors and shades, anticipating modern color lithography by at least sixty years. All the shades of one color were printed from one stone, whereon a parallel line screen of uniform density had been drawn mechanically and subsequently etched (fig. 6.4).

Wet-collodion plate negatives of the complete line drawing of a map are produced by the camera.

Stones are coated (in the dark) with a solution of silver nitrate.

The glass negative is printed on the light-sensitive surface of the stone by exposure to sunlight (in the sun, twenty-four hours; in winter, with covered sky, about six days).

Developing and fixing (about fifteen hours).

The stone is coated with an acid-resistant coating of wax and soluble bitumen.

The transparent hardened protective coating is placed under an engraving machine, which covers the whole surface of the coating stone with parallel grooves, cut by a diamond, from eight to twenty lines to a millimeter.

Those parts of the map to appear white are covered with the acid-resistant coating of bitumen.

Parallel line screens are etched into the surface of the stone by a selective process in four to six phases. In the first phase, the full area is etched during one minute, producing the lightest shade.

The areas with lightest shades are covered with acid-resistant lithographic ink. After the lithographic ink has dried, the second shade is obtained by exposing the stone to acid of the same strength, but one minute longer.

The areas of the second shade are covered with lithographic ink and the etching is repeated for the other shades, and so on. By this repeated etching process, the darkest shade has been etched for about eight minutes, by which grooves of considerable thickness are produced in the lithographic stone.

Finally, the stone is cleaned with turpentine. Supplementary stone engraving by hand is required for those lines that have to be printed in the same color as the shades.

Map sheets are printed from four stones: black, yellow, red, and blue. A special frame with pins to guarantee perfect register is attached to the printing press. All the sheets of printing paper are punched beforehand—presumably the earliest example of register-punching in cartography.

During the period 1864–1910, the following maps reproduced by the Eckstein process were published:

A series of twenty-two topographical maps of the residencies on Java and Madura, scale 1:100,000;

The "Waterstaatskaart" of the Netherlands, 1:50,000, in 175 sheets;

The chromo-topographical maps of the Netherlands, 1:25,000, only for the green plate, showing meadows and forests;

The hypsographic map of the Netherlands, 1:600,000, with layer tints;

Post routes of the Netherlands, 1:250,000;

Topographical map of Atjeh, Sumatra, 1:40,000.

The topographical maps of Java and Madura, in particular, were admired in and outside the Netherlands. Their complex contents of land use, vegetation, and relief required color reproduction in various tints. As early as 1865 it was decided that their reproduction should be undertaken by the Topographic Department of the Ministry of Defence at the Hague.[11] Another big project was the reproduction in color of the "Waterstaatskaart," 1:50,000, a map showing the distribution of polders and polder authorities by colors and tints of colors.[12] The map was initiated in 1864 and the first edition of 175 map sheets was executed in delicate colors, superimposed on a violet topographical base. (fig. 6.5).

Various military and private printing industries showed great interest in the Eckstein process. It was applied in Sweden for the reproduction of the geological map, a description and a specimen of which is given in a Swedish paper by Börtzell of 1872.[13] It was also used by the Austrian Military Survey for some special colored maps. Because of its time-consuming and expensive platemaking process the Eckstein method was gradually given up and replaced by less costly but less beautiful reproduction methods. On many occasions, maps by the Eckstein process were shown at international exhibitions. In 1866 it was awarded a gold medal at Amsterdam (Eckstein was then technical director of the Topographic Bureau; Colonel Besier was general director). A special publication with maps and plates accompanied their stand at the International Exhibition at Philadelphia in 1876.[14] The well-known German cartographer Carl Vogel reported about the world exhibition at Paris in 1878 as follows:

was die Ausstellung des Niederländischen Generalstabs noch besonders interessant machte, das war die Ansicht einzelner Steine von welchen die Karten gedruckt waren und die das neue Chromo-lithographische Verfahren veranschaulichen sollten, von welchen oben die Rede war. Eben so lag eine Probe von Farbendruck aus, wonach man von nur 3 Steinen durch Anwendung der Ätzmethode (procédé Eckstein) 18 verschiedene Nuancen derselben Farbe erhält.[15]

Comparing the Eckstein method with other color printing processes makes it clear that here half-tone screens have been used in the proper way, by superimposing screens of the primary colors to obtain other colors. This does not mean that the Eckstein method was the only process to print colors by means of patterns. When looking at colored atlas and map sheets of the years around 1880, one often observes certain textures in the flat colors. These textures or patterns were transferred to the stone from prefabricated paper sheets. (Compare our modern preprinted shading film or rub-on patterns.) At that time, geological maps were much in demand. Conventional symbols for geological maps had to be printed in color, in shades of color, or in patterns. It is interesting to study the advanced state of prefabricated patterns used for the first geological maps of the United States Geological Survey and laid down in the "Standards for Geological Cartography" in 1889.[16]

Two disadvantages caused the demise of the Eckstein method: revision of the plates was extremely difficult, and the process was very time consuming. But it leaves us with one of the most attractive map specimens of the nineteenth century.

Heliogravure, combined with electroplating. In the course of the second half of the nineteenth century, various processes were invented to create an intaglio printing surface by means of photochemical etching (see the chapter by Elizabeth M. Harris). In France, Poitevin produced a galvanoplastic copper plate from a gelatine mold. In England, Edwards and also Toovey used so-called zincography, in which process a zinc plate was etched under a protective coating. In Bavaria, "Albertypie," a process named after the staff officer Albert, was used for the printing of short runs from a hardened gelatine surface in relief, with a thick glass plate as base. In 1854 Paul Pretsch of Vienna invented the process for making intaglio plates from swelled-gelatine prints. He used bichromated gelatine, exposed under a diapositive (in the sun or in daylight, of course). After swelling the non-exposed gelatine with cold water, a galvanoplastic process produced an intaglio copperplate. Similar methods have been more or less successfully used. In England W. B. Woodbury (1870) removed the developed gelatine coating from the glass plate. The gelatine film was dried, hardened, and pressed under a hydraulic press between a steel and lead plate. Thus an impression was produced in the softer metal, which, after hardening, was used for printing. (In France, the Woodbury type was known as *Procédé Photoglyptic.*) This process may have been inspired by the so-called autoplastic or *Natur-selbstdruck,* invented in 1851 by the director of the Austrian government printing house, Alois Auer. He achieved very remarkable results by pressing such objects as plant leaves or flowers into a lead plate, which was copied afterward by an electrotyping process.

Electroplating, producing an electrotype, makes use of a wax mold from the printing plate. This mold is rendered electrically conductive by powdering it with graphite. Placed in a tank with a solution of copper sulphate, the mold serves as cathode, while a plate of copper serves as anode. The copper is deposited on the mold, forming a copper printing plate.

In Vienna, heliogravure brought very great fame to map reproduction. Between 1872 and 1885 the K.u.K. Militär-Geographisches Institut amazed the world and was praised throughout Europe for its unparalleled achievement in topographic mapping. It not only made use of photography for the reduction of the fair drawing at the scale of 1:60,000 to the reproduction scale 1:75,000 but also applied an electroplating process for the production of intaglio plates. In all, 715 sheets were produced in twelve years. The quality of the prints was excellent and could match that of copper engraving. Actually, this printing process was very similar to copperplate printing, apart from the fact that the copperplate was not engraved by hand, but by heliogravure and electroplating. (The electrolytic deposition of a copperplate took four to six weeks).[17]

In an article, W. F. Versteeg tried to influence the Dutch authorities to give up lithography and to revert to intaglio printing by heliogravure platemaking.[18] Throughout the nineteenth century, the competition between copper engraving and lithography stimulated numerous discussions by cartographers. As early as 1826, E. F. Jomard raised the point whether lithography was preferable to copper engraving for the reproduction of maps.[19] About 1870, there was still a preference for the superior graphic quality of the copper engraving, in particular among the older generation. But the mapmaker was forced to use stone because of the growing shortage of engravers and because the process was becoming increasingly uneconomic.

In a letter to Augustus Petermann, dated 19 February 1878, the director of the Austrian Military Survey gives the following comparison in time and cost between heliogravure and copper engraving as an average for a map sheet 28 × 27 cm.:

Heliogravure		*Copper engraving*	
drawing	7.5 months	drawing	3.7 months
reproduction	1.5 months	engraving	42.0 months
costs, drawing	1,000 fl.	costs, drawing	500 fl.
costs, plates	45 fl.	costs, engraving	3,500 fl.
costs, retouching	50 fl.	costs, plate	140 fl.[20]

From this, copper engraving appears to be approximately four times as expensive. But for us, in our time, it is amazing to learn from the same source as cited above that the heliogravure process of exposure, developing, electroplating, and retouching took about five weeks for one plate.

Fifty years after the heyday of Viennese heliogravure, the process was tried out again by the military survey in Berlin. In the Reichsamt für Landesaufnahme, about 1930, a new process of platemaking was introduced, which was actually the same as that used in Vienna in 1870–80, but much improved by powerful lighting.[21] The final product was a set of three intaglio copperplates for printing the three colors black, brown (for contour lines), and blue (water) of the German topographical map 1:25,000. Although about 1930 offset printing from zinc plates had proved itself very suitable for map reproduction, a powerful pressure group in Germany still preferred intaglio engraving. Their influence in the governmental cartographic organizations was very strong, which led to an archaic but very beautiful map reproduction method in the period 1930–40.

Production speed in map printing. Obviously, the quantity of military topographical maps needed for immediate use was very small. In the Austrian Military Survey, fifteen copperplate presses were in use to print copies from the 715 plates, each press printing six to eight copies an hour. The lithographic press was considerably faster; the hand press produced twenty to fifty copies an hour and the steam driven high-speed press produced three to five hundred copies an hour.[22] The fact is often ignored that the stone on the printing press was *not* the stone on which the map had been drawn or engraved. For various reasons (faster inking, risk of damaging the original) the image was transferred to the printing stone, where it remained intact for two to four thousand prints. After that, a new transfer had to be carried out.

The Collotype printing process.[23] Invented in 1869, long before the application of arc light lamps, this process brought the solution of the problem of reproducing continuous-tone documents by means of the printing press. Although not of direct importance for map reproduction, it deserves mention, be it only for the printing of some beautiful early hill shading in atlases.[24] It is a surface-printing process of great beauty and delicacy, but also comparatively slow, and therefore most suitable for small editions of fine quality.

A plate, usually glass, is coated with gelatine and a bichromate and exposed to light under a continuous tone negative. After a series of developing processes with water and glycerine the gelatine surface is ready to receive ink. The reception of ink by the gelatine is controlled by the amount of moisture present in its reticulations. The various tones of the negative are perfectly reproduced. This screenless reproduction process shows a perfection in detail and a softness of tone unequaled by any other process.

Photolithographic printing after 1900. Two fundamental inventions brought a revolution in map reproduction during the early years of the twentieth century: the invention of the chromium-gum copying process on zinc by Dr. Strecker in 1901, and the invention of the offset press in 1904.

1900–40

The chromium-gum copying process. Before 1900, various experiments were carried out using zinc or aluminum as a printing surface.[25] The problem was to find a light-sensitive coating strong enough to allow for a large edition. During the exposure, the emulsion hardens and, after developing, takes the ink, but in course of time the printing elements wear out. Dr. Strecker invented a new platemaking method, still in use, called the deep-etch method. Instead of a negative, a positive is printed down on the chromium-gum coating. After exposure, the non-hardened lines under the diapositive are washed off and the bare metal is etched with a weak acid. The hardened emulsion under the transparent parts of the diapositive serves as a protective coating. Next the plate is rubbed with a lacquer that sticks to the metal. Subsequently a greasy ink, which sticks to the lacquer, is rubbed on the plate. It is then cleaned with brush and water and is ready for printing eight to ten thousand copies.

Now the printers waited for a press whose speed could match the almost everlasting durability of the plate. The invention was first announced in *Klimsch Jahrbuch* 3 (1902), but it took several years before flat-bed printing from zinc or aluminum became general practice. Both research and first application were carried out in the laboratories of the Klimsch Firm in Frankfurt-am-Main. In 1905, in one of the best early articles on map reproduction, F. Hesse wrote: "'der Firma Klimsch ist es nun nach mancherlei Versuchen gelungen, ein Positiv Kopierverfahren ausfindig zu machen.'"[26]

Before 1900, printing from zinc or aluminum was tried out, but without satisfactory results. Small numbers of copies could be produced from plates made with copy negatives, but the quality of longer runs was poor. In most of the topographical bureaus in and outside Europe, printing from stone or copperplate was preferred, even when the new Strecker process proved that it could compete with the old process. Not until 1925–30 had most of the national topographical surveys switched over from stone to zinc. This is one of the reasons why the offset printing press came into use so late.

Mention has been made of the negative platemaking process. This process, first applied around 1880, was developed into an excellent platemaking method, nowadays generally used in the United States and in many of the English-speaking countries. A negative is printed on a light-sensitive emulsion of albumen and chromium. The emulsion hardens under the exposed lines of the negative and the non-exposed area is washed off after the plate has been developed with a greasy developing ink. A characteristic difference between the positive and negative process is that plates produced in the positive process last longer so that a larger number of copies can be produced from one plate.

The offset printing press. In 1904, the American Ira W. Rubel is presumed to have invented the offset process. His invention was made by accident, when a sheet of paper failed to pass between the flat plate and the pressure cylinder. The next sheet got a mirror-reversed impression from the cylinder on its reverse. When Rubel applied for a patent, he was surprised to hear that the offset process already existed. It had been in use since 1876, when Barclay had obtained a patent for printing on a metal surface to decorate metal boxes. In France, Voirin constructed in 1879 a flatbed offset machine with a rubber blanket for printing on paper as well as on metal sheets. The same Voirin constructed the first rotating offset machine in 1910. In order to profit from the much higher speed, however, the paper feed had to be automated.

The principle of the offset press rests on the transfer of the ink image to an intermediate surface: a resilient rubber blanket mounted on a cylinder, which "sets off" the ink image on the paper. The zinc or aluminum printing plate is attached to a rotating cylinder that receives ink and water alternately. Unlike that of a flat-bed press, the paper is not in direct contact with the wet plate, thus reducing the amount of moisture on the paper sheets.

European map producers were not at all impressed by the high speed of the rotary press, because European editions were relatively small.[27] In Austria, the Military Topographical Survey was the first to install an offset press in 1910, soon followed by two others. In the United States, however, the offset press fulfilled a long-entertained desire to increase the production speed. Early offset presses produced about fifteen hundred prints an hour. Modern offset presses run seven to eight thousand prints an hour, a speed seldom required for map reproduction. In addition, from the viewpoint of quality control, it is usually not recommended to exceed four thousand copies an hour.

Two-color offset presses came into use after 1910, but in Europe their application in map printing was restricted to the military survey and its use in civil mapping was seldom seen before 1945. One reason why the offset press was accepted so slowly among the map printing houses in Europe was the huge stock of original stones. Most of the map series were tied to the printing surface of the stone. Only when new maps had to be drawn could one consider putting them on zinc.

The decline of cartography in the first half of the twentieth century. In sharp contrast to the technical perfection connected with repro-cameras and offset presses stood the recession of cartographic drawing in the first half of the twentieth century. This conflicting situation was the all-absorbing topic of professional talk, mainly after 1925. Various causes had led to this most unhappy position. The main root of all the trouble was the introduction of transparent plastics in map drawing. Initially, celluloid has to be held responsible for lowering the standard of map drawing. Celluloid, invented in 1865 by Alexander Parkes of Birmingham and first manufactured by the Hyat brothers in America in 1872, greatly contributed to the development of conventional photography. However, as a base for emulsions, used for map reproduction, it was unfit because of its dimensional instability. After 1908, a product named Zellon (cellulose nitrate) was often used for drawing with pen and ink. Advantages of transparency and ease of handling led to the sacrifice of graphic quality. Between 1908 and 1930, innumerable maps were spoiled by using Zellon as a drawing base.[28] Next came the acetates: Kodatrace, etc. Thanks to a slightly better dimensional stability, hosts of qualified draftsmen were forced to waste their skill by drawing on a surface hostile to ink. Imperfect ink cover and defective lines gave imperfect printing plates and resulted in poor impressions. Inconsistent line weight and coarse line elements spoiled the traditional graphic beauty of the copper engraving or lithographic drawing.

A second factor leading to the decline of cartography was of a more general kind: a fast-increasing demand for maps in a world wherein mankind was carried away by technical progress. Handwork was replaced by mechanization and speed was required. By 1930, the generation of copper engravers had almost died out. In the drawing offices of survey departments, the time-consuming engraving on copper or stone was abandoned. The clumsy stones were replaced by acetate sheets and the unparalleled skill of the copper or stone engraver was no longer needed.

A third reason for the lowering of cartographic standards may be seen in the introduction of aerial photogrammetry. New editions of topographical maps, based on the more reliable aerial photographs, offered the opportunity to start the map-making process anew. In view of the speed of the aerial survey, cartographic drawing had to speed up as well, in order that it should not function as a bottleneck in the flow of the mapping process. Whenever speed is required, the graphic quality must suffer.

Bright spots in this dark report are the private cartographic institutes in Europe still following the great traditions, like John Bartholomew and Son in Edinburgh and Justus Perthes in Gotha. In one instance, a topographical survey reverted to intaglio printing from copperplates produced by means of electroplating: the Reichsamt für Landesaufnahme, Berlin, 1935.[29] Notwithstanding the perfection of transparent plastics by the production of polyvinylchloride sheets since about 1940 and polyester sheets after 1950, drawing with special inks on those drawing surfaces did not bring back the "Paradise Lost" of nineteenth-century map quality.

Cartography was in this deplorable state on the eve of the second revolution in map reproduction. The first revolution was characterized by the introduction of photography, toward the middle of the nineteenth century. The second revolution began around 1940. As in so many disciplines and technologies, cartography has undergone more changes in the last twenty-five years than in centuries before.

The restoration of the cartographer's craft. The tragic struggle of the draftsman with his ink and plastic ended when cartography reverted to engraving again, no longer on metal plates but on the ideal surface of the glass plate or its substitute: the transparent plastic sheet, coated with a thin, soft film, opaque to actinic light, so that a negative was produced. Needles of steel or sapphire, sharpened to required line weight, afforded identical line width for each map element. Constant force, pressure, and direction are secured by toolholders. The name of the new technique was scribing, but the result could well match the old copper or stone engraving.

Like many techniques, scribing on coated glass plates or transparent plastic sheets was invented several times. The artists Corot and Daubigny already had engraved their graphic work on glass in 1854.[30] One of the oldest recipes for a scribe-coat emulsion was in an 1884 manual of industrial arts, described by David Woodward.[31]

It is difficult to find out where scribing first was used. The Swiss Federal Topographical Surveys were using scribing on glass on a production base for the topographical map series about 1945. In the Soviet Union, scribing on glass has been in use since about 1937.[32] In Switzerland, the town plans of Zurich were engraved on coated Zellon in 1933. Various topographical bureaus have reminiscences going back to the nineteenth century of map work for which the color separation of the line elements (e.g., contours) was executed by means of scribing in coatings on a glass plate. Contrary to what has been suggested in literature, it was neither in Europe nor in America that scribing was first used on a production base, but in the Netherlands East Indies. In 1927, at the Topographic Surveys in Batavia, scribing on coated glass plates was introduced, not experimentally, but as a working method for the fair drawing of series topographical maps at various scales (fig. 6.6).[33] This very successful method remained in use until the capture of the East Indies by the Japanese in 1942. One of the famous works produced by this method is the *Atlas van Tropisch Nederland* (1939). The scribe-coating, consisting of a solution of bitumen in turpentine, was applied to glass negatives. Scribed negatives were produced for the separation of colors in the printing process. The introduction of scribing on coated plastic in Europe was due mainly to the efforts of federal mapping agencies in the United States. Since 1940, the Coast and Geodetic Survey has applied scribing to the production of various map series. After 1945, Europe gradually became informed about the advantages of this new technique, and one of the first to use it was the Generalstabens Litografiska Anstalt at Stockholm, Sweden, in 1947.

1940–70   To label the period 1940–70 as the second revolution in map reproduction, there is more required than simply a new manual drawing technique. When looking at the last thirty years we observe the following fundamental new developments:

The manufacturing of plastic sheets, adapted to the requirements of cartography.

Strip-masking techniques for applying gradient tints and colors to plastic sheets. Gradient tints were formerly produced by manually opaquing selected areas on blue-line copies. In the new process, open lines are produced by photochemical etching in an opaque coating, which is easily peeled off, by which the sheet is rendered into an open-window negative.

The use of film screens for copying tints and patterns.

The application of prefabricated stick-up lettering, screens, and symbols.

Photo composition of map lettering. Text and names on maps previously were drawn freehand or were composed from type and printed separately. Photo composition produces diapositives of photographed type at any required size. These names are stuck on to the lettering film.

Color-proofing on plastic sheets (the so-called Hausleitner copying process).

By combining the six new techniques we arrive at a purely photomechanical map compilation process wherein the traditional manual skill required of the draftsman is reduced to a minimum, but wherein the traditional high level of graphic quality is guaranteed. Summarizing, we can conclude that photomechanical map reproduction is expedient for cartography in a period when more maps have to be produced by less qualified draftsmen. The transitional period 1940–60, when drawing paper was replaced by plastic sheets, when draftsmen were converted into new-style engravers, when map compilation became dependent on dexterity with bottles, brushes, cutting tools, glue, and zip-a-tone, when the cartographer's outfit became more like a chemist's shop, created many deep-felt emotions among the old-style cartographers. These emotions are reflected by the literature of the period 1950–60 in European journals.[34]

In the domain of photography, some very interesting and important new applications were seen after 1950. Mention should be made of the new designs of repro-cameras, the perfection of the mercury vapor and xenon lamp, the application of diazo materials, pre-sensitized plates, polyvinyl-alcohol as a colloid, and the progress of electrostatic printing.[35]

Bewilderment and doubt about the present state and future of map reproduction techniques were steered into calmer channels by international conferences of experts on cartography. Special mention should be made of the first conference of this kind in Stockholm in 1956, the so-called ESSELTE Conference.[36] One of the most important conclusions drawn from the developments in the last two decades is that the main process of map reproduction has been transferred from lithography to photography. In other words, many operations in the map compilation and plate-making process have been shifted to the photographic section of the cartographic institute. This field was opened by the improvement of repro-cameras and of photographic materials and chemicals. Summarizing this development, one arrives at the conclusion that the fundamentals of photoreproduction and lithographic printing have not changed so much but that a very high level of quality was achieved, enabling the demands of a map-swallowing society to be met.

It seems quite appropriate to conclude this lecture by referring to the Second International Cartographic Conference, held near Chicago, in Evanston, under the auspices of Rand McNally and Company. Its proceedings reflect the stabilization and the acceptance of photo-mechanical and photochemical processes by the leading map producing organizations in the United States and Europe.[37] The proceedings also anticipated the birth of a new area in cartography: automation. That automation also affected cartography like any other manual technique is reflected by the vast amount of literature (over three thousand items in ten years) on the subject.

In the era of automation there will be little room for human aspects of the map-maker's craft. Thus, the contents of the Chicago Cartographic Conference of 1958 are counterbalanced by this Chicago conference on the history of map printing, where human abilities have manifested themselves so well.

Notes

1. Mapmaking and map printing: The evolution of a working relationship

1. The psychological and intellectual effects of the development of printing have been largely limited to those that stemmed from movable type, while the significant effects of the duplication of drawings and other graphics generally have been ignored, although Ivins (William M. Ivins, Jr., *Prints and Visual Communication* (Cambridge, Mass.: Harvard University Press, 1953 [M. I. T. Press paperback]) is a notable exception. An interesting example of the effects of both is touched on in Peter H. Raven, Brent Berlin, and Dennis E. Breedlove, "The Origins of Taxonomy," *Science* 174 (1971): 1210–13.

2. R. A. Skelton, "The Early Map Printer and His Problems," *The Penrose Annual,* ed. by Herbert Spencer (London: Percy Lund, Humphries & Co., Ltd., 1964, pp. 170–86 [ref., p. 171]).

3. Ivins, *Prints and Visual Communication,* p. 16.

4. David Woodward, "Hand-Engraving, Movable Type and Stereotyping: A Study of the Lettering Methods for Woodcut Maps," Master's thesis, University of Wisconsin (Madison), 1967, pp. 20–24.

5. Kenneth Nebenzahl, "A Stone Thrown at the Map Maker," *The Papers of the Bibliographic Society of America* 55 (1961): 283–88.

6. Arthur M. Hind, *An Introduction to a History of Woodcut,* 2 vols. (Boston: Houghton Mifflin, 1935 [Dover paperback, 1963]), p. 1.

7. Cor. Koeman has called to my attention the entry in J. G. Krünitz, *Oekonomsch-Technologische Encyklopaedia* (Berlin, 1793), vol. 60, pp. 152–60, which states that a copperplate could produce 5,000 prints at a maximum while a woodblock could produce over 200,000. Steel plates, softened by heat so that they could be worked with engraving tools and afterward hardened, were tried in the early nineteenth century in France and England in order to obtain a longer printing run. They were harder to correct and apparently the process was not much used. See R. A. Skelton, "Cartography," *A History of Technology,* Charles Singer et al., eds. (New York: Oxford University Press, 1958), 4:596–628.

8. Michael Clapham, "Printing," *A History of Technology,* Charles Singer et al., eds. (New York: Oxford University Press, 1957), 3:377–416.

9. Woodward, "Hand-Engraving," passim.

10. David Woodward, "Some Evidence for the Use of Stereotyping on Peter Apian's World Map of 1530," *Imago Mundi* 24 (1970): 43–48.

11. R. A. Skelton, *Maps, A Historical Survey of Their Study and Collecting* (Chicago: University of Chicago Press, 1972), p. 16.

12. G. Walters, "Engraved Maps from the English Topographies c. 1660–1825," *Cartographic Journal* 7:2 (1970): 81–88. See also R. V. Tooley, *Maps and Map-Makers,* 2d ed. (New York: Crown Publishers, Bonanza Books, 1961), pp. 65–66.

13. John Temple Leader, *Life of Sir Robert Dudley Earl of Warwick and Duke of Northumberland* (Florence: G. Barbera, 1895), p. 122.

14. Joseph Amann, *Die bayerische Landesvermessung in ihrer geschichtlicher Entwicklung. Erster Teil: Die Aufstellung des Landes vermessungwerkes, 1808–1871* (Munich: K. B. Katasterbureau, 1908), pp. 125–26.

15. Leader, *Life of Sir Robert Dudley,* p. 122.

16. Skelton, "Cartography," p. 625.

17. Arthur M. Hind, *A History of Engraving and Etching from the 15th Century to the Year 1914* (Boston: Houghton Mifflin, 1923 [Dover paperback, 1963]), p. 13.

18. Arthur H. Robinson, "The 1837 Maps of Henry Drury Harness," *Geographical Journal* 121 (1955): 440–450. For a description of the aquatint process see S. T. Prideaux, *Aquatint Engraving* (London: Duckworth and Co., 1909).

19. Amann, *Bayerische Landesvermessung,* p. 127.

20. Alois Senefelder, *The Invention of Lithography,* trans. from the German by J. W. Muller (New York: Fuch and Lang Mfg. Co., 1911), pp. 98–99.

21. M. Jomard, "Examen de la question de savoir: 1. Si la lithographie peut-être appliqué avec avantage à la publication des cartes géographiques, tant sous le rapport du mérite de l'exécution que sous le rapport de l'économie? 2. Jusqu'à quel point elle peut remplacer pour cet objet la gravure sur cuivre?" *Bulletin de la Société de Géographie de Paris* 1. série, 5(1826): 693–708, including a table at end of volume.

22. Alois Senefelder, *A Complete Course of Lithography Containing Clear and Explicit Instructions in all the Different Branches and Manners of that Art: . . . a History of Lithography . . . ,* trans. by A. S. (London: R. Ackermann, 1819), pp. 80–85.

23. Ibid., p. 256.

24. Ibid., p. 281.

25. David Woodward, "Cerotyping and the Rise of Modern American Commercial Cartography, A Case Study of the Influence of Printing Technology on Cartographic Style," Ph.D. dissertation, University of Wisconsin (Madison), 1970.

26. S. T. Prideaux, *Aquatint Engraving . . .* (London: Duckworth, 1909), pp. 32 ff.

27. Clapham, "Printing," *History of Technology,* 3:377–411.

28. F. Grenacher, "The Woodcut Map," *Imago Mundi* 24 (1970): 31–41.

29. Joseph Fischer, S.J., and Father R. V. Weiser (eds.), *The Oldest Map with the Name America of the Year 1507 and the Carta Marina of the year 1516 by M. Waldseemüller (Ilacomilus)* (Amsterdam: Theatrum Orbis Terrarum, Ltd., 1968 [reprint]), p. 4.

30. A. S. Osley, *Mercator: a monograph on the lettering of maps . . .* (New York: Watson-Guptill Publications, 1969).

31. John Arlott (ed.), *John Speed's England, A Coloured Facsimile of the Map and Text from the Theatre of the Empire of Great Britaine, First Edition, 1611, Part I* (London: Phoenix House Ltd., 1953), p. 7.

32. C. Koeman, *Joan Blaeu and his Grand Atlas* (Amsterdam: Theatrum Orbis Terrarum, Ltd., 1970), p. 18.

33. Charles Close, *The Early Years of the Ordnance Survey,* with a new introduction by J. B. Harley (Newton Abbot, Devon: David and Charles Reprints, 1969), pp. 71–72, 127.

34. Ibid., p. 126.

35. Gerhard Engelmann, "Der Physikalische Atlas des Heinrich Berghaus und Alexander Keith Johnstons Physical Atlas," *Petermanns Geographische Mitteilungen* 108 (1964): 133–49.

36. Gerhard Engelmann, "Carl Ritters 'SECHS KARTEN VON EUROPA,' " *Erdkunde* 20 (1966): 104–10. *See also* K. Sinnhuber, "Carl Ritter 1779–1859," *Scottish Geographical Magazine* 75:3 (1959): 152–63.

37. Carl Ritter, nos. 11 and 12 in "Sammelband mit 24 Karten Carl Ritters," a portfolio of manuscript materials in Kartenabteilung der Staatsbibliothek, Marburg an der Lahn.

38. A reduced printing of the map is included in Edgar Lehmann, "Carl Ritters kartographische Leistung," *Die Erde* 90 (1959): 184–222 (reproduction on p. 190).

39. Engelmann, "Carl Ritters 'SECHS KARTEN,' " p. 106.

40. Douglas C. McMurtrie, *Printing Geographic Maps with Movable Types* (New York: privately printed, 1925).

41. Woodward, "Cerotyping," and Charles Eckstein, *New Method for Reproducing Maps and Drawings,* Topographical Department, War Office [Netherlands], and Giunta d'Albani Bros., The Hague (International Exhibition at Philadelphia, 1876). Eckstein described, for lithography, the production of graded tones, the three-color process, and the use of movable type.

42. William Gamble, *Line Photo-Engraving* (London: Percy Lund, Humphries & Co., Ltd., n. d.), pp. 9–13, and Ivins, *Prints and Visual Communication,* pp. 113–34.

43. Otto C. Stoessel, *Standard Printing Screen System,* technical report 72–1 (Saint Louis: Aeronautical Chart and Information Center, 1972).

44. Ernst Spiess and Max Buhlmann, *Zur Aufrasterung von Strichvorlagen* (Zurich: Kartographisches Institut, Eidgenosse Technische Hochschule, 1971).

2. The woodcut technique

1. Arthur M. Hind, *An Introduction to a History of Woodcut,* 2 vols. (Boston: Houghton Mifflin, 1935; reprinted New York: Dover Publications, 1963), p. 1.

2. Thomas Francis Carter, *The Invention of Printing in China and its Spread Westward,* 2d ed., revised by L. Carrington Goodrich (New York: Ronald Press, 1955), chaps. 7 and 21.

3. Hind, *Woodcut,* p. 96.

4. S. H. Steinberg, *Five Hundred Years of Printing,* 2d ed. (London: Penguin Books, 1961), p. 158.

5. Isidorus Hispanensis [Saint Isidore of Seville], *Etymologiarum sive Originum libri XX* (Augsburg: Günther Zainer, 1472), fol. 177v.

6. *Rudimentum novitiorum* (Lübeck: Lucas Brandiss, 1475).

7. Robertus Valturius, *De re militari* (Verona: Johannes Nicolai de Verona, 1472). See *Printing and the Mind of Man* (London: F. W. Bridges, et al., 1963), p. 33.

8. A. Hyatt King, *Four Hundred Years of Music Printing* (London: Trustees of the British Museum, 1968), p. 9.

9. Franz Grenacher, "The Woodcut Map," *Imago Mundi* 24 (1970): 31.

10. For example, in Johann Stumpf, *Schwyzer Chronik* (Zürich, 1606).

11. For example, in *Articles of Agreement made and concluded upon between the Right Honourable the Lord Propietary (sic) of Maryland and The Honourable the Proprietarys of*

Pensilvania, &c, . . . (Philadelphia: B. Franklin, 1733), and in *Gentleman's Magazine*, 11 (1742): 445.

12. J. R. M. Papillon, *Traité Historique et Pratique de la Gravure en Bois*, 3 vols. (Paris: Pierre Guillaume Simon, 1766).

13. Hind, *Woodcut*, pp. 1–28.

14. Ibid., pp. 3, 9.

15. This reference has not been authenticated. See ibid., pp. 208–9.

16. Papillon, *Traité* 2:64–74.

17. Ibid., 2:12.

18. Ibid., 2:29–32.

19. Hind, *Woodcut*, p. 10.

20. Alexander Nesbitt, *The History and Technique of Lettering* (New York: Dover, 1957), pp. 50–54.

21. Johann Gottlob Immanuel Breitkopf, *Ueber den Druck der Geographischen Charten, Nebst beygefügter Probe einer durch die Buchdruckerkunst gesetzen und gedruckten Landcharte* (Leipzig: J. G. I. Breitkopf, 1777), p. 9.

22. William Andrew Chatto, *A Treatise on Wood Engraving, Historical and Practical,* 2d ed. (London: C. Knight, 1861), p. 247. Justin Winsor, *A Bibliography of Ptolemy's Geography.* Bibliographical Contributions, no. 18. (Cambridge, Mass.: Harvard University Press, 1884), p. 8. Henry N. Stevens, *Ptolemy's Georgraphy. A brief account of all the printed editions down to 1730 . . . ,* 2d ed. (London: Henry Stevens, Son & Stiles, 1908). Thomas W. Huck, "The Earliest Printed Maps," *The Antiquary* 46 (1910): 255. Henry N. Stevens, *The First Delineation of the New World and the first use of the Name America on a Printed Map,* (London: Henry Stevens, Son & Stiles, 1928), pp. 7, 9–12, 69, 83, 109. Jose Anesi, "Aspectos Graficos de la Cartografia," *Revista Geográfica Americana* 30 (1948): 169–84. François de Dainville, *Le Langage des Géographes,* (Paris: A. & J. Picard, 1964), p. 70. R. A. Skelton, bibliographical note in *Claudius Ptolemaeus: Geographia* (Venice, 1511), p. x (Amsterdam: Theatrum Orbis Terrarum, 1969).

23. Stevens, *New World,* p. 69.

24. Simon Grynaeus, *Novus orbis regionum ac insularum veteribus incognitarum . . .* (Basel: H. Petri, 1532).

25. Henrich Bünting, *Itinerarium et chronicon ecclesiasticum totius Sacrae Scripturae,* 2 vols. (Magdeburg: A. Duncker, 1597).

26. Chatto, *Wood Engraving,* p. 247.

27. Otto Hupp, *Philipp Apian's Bayerische Landtaflen und Peter Weiner's Chorographia Bavariae, Eine Bibliographische Untersuchung,* (Frankfurt-am-Main: Heinrich Keller, 1910), p. 12.

28. Douglas C. McMurtrie, *Stereotyping in Bavaria in the Sixteenth Century—A Note on the History of Map Printing Processes and of Printer's Platemaking* (New York: privately printed, 1935).

29. Johann Christian Freiheit von Aretin, "Die Stereotypen in Baiern im XVI Jahrhundert erfunden," *Beyträge zur Geschichte und Literatur* 2 (1804): 72.

30. Hupp, *Apians Landtafeln,* 13.

31. See, for example, the following: Huck, "Earliest Printed Maps"; Siegmund Günther, "Peter und Philipp Apian, zwei deutsche Mathematiker und Kartographen," *Abhandlungen der Königliche böhmischen Gesellschaft der Wissenschaften* 6 (1882): 121; R. A. Skelton, *Decorative Printed Maps of the 15th to 18th Centuries* (London: Staples Press, 1952), p. 14; G. R. Crone, "The Evolution of Map Printing," *Printing, A Supplement, issued by "The Times" on the occasion of the 10th International Printing Machinery and Allied Trades Exhibition, July 1955;* Margaret Oldfield, "Five Centuries of Map Printing," *The British Printer* 63 (1955): 69–74.

32. David Woodward, "Some Evidence for the Use of Stereotyping on Peter Apian's World Map of 1530," *Imago Mundi* 24 (1970): 43–48.

33. Hupp, *Apian's Landkarten,* and Werner Horn, "Sebastian Münster's Map of Prussia and the Variants of It," *Imago Mundi* 7 (1951): 66–73.

34. Karl Heinz Burmeister, *Briefe Sebastian Münsters, Lateinisch und Deutsch* (Frankfurt-am-Main: Insel-Verlag, 1964).

35. Chatto, *Wood Engraving,* p. 497.

36. W. Turner Berry and H. Edmund Poole, *Annals of Printing* (London: Blandford Press, 1966), p. 182.

37. David Woodward, "The Foster Woodcut Map Controversy: A Further Examination of the Evidence," *Imago Mundi* 21 (1967): 51–61.

38. Hind, *Woodcut*, pp. 2–6.

39. Ibid., pp. 2–3.

40. R. A. Skelton, "Colour in Mapmaking," *Geographical Magazine* 32 (1960): 544–53.

41. Chatto, *Wood Engraving.*

42. Hind, *Woodcut*, pp. 168–69.

43. Grenacher, "Woodcut Map," pp. 31–41.

3. Copperplate Printing

1. T. C. Hansard, *Typographia . . .* (London, 1825), p. 22. Hansard states that this was discovered about 1450 by Maso Finiguerra, a goldsmith.

2. British Museum, *A Guide to the Processes and Schools of Engraving* . . . (London, 1914), p. 20.

3. Leo Bagrow and R. A. Skelton, *History of Cartography* (London, 1964), p. 91.

4. The basic references for this paper are as follows: William Faithorne, *The Art of Graveing and Etching* . . . , 2d ed. (London: A Roper, 1702). This edition contains additional material on copperplate engraving and printing not in the first edition of 1662. *Account of the Methods and Processes Adopted for the Production of the Maps of the Ordnance Survey*, 2d ed. (London, 1902).

5. Faithorne, *Graveing*, p. 4.

6. R. H. Brown, "The American Geographies of Jedediah Morse," *Annals, Association of American Geographers* 31 (1941): 175.

7. British Museum, Maps 55.b.31.

8. A. Arrowsmith, *Memoir Relative to the Construction of The Map of Scotland* (London, [1809]), p. 31.

9. Abraham Bosse, *Traicté des Manières de Graver* (Paris, 1645).

10. Coolie Verner, "Mr. Jefferson Makes A Map," *Imago Mundi* 14 (1960): 102.

11. David Woodward, "Seventeenth Century Copper Engraving for Maps" (unpublished).

12. *Account*, p. 166.

13. Coolie Verner, *Smith's Virginia and Its Derivatives*. Map Collector's Circle No. 45 (London, 1968), p. 38.

14. Coolie Verner, *Captain Collins' Coasting Pilot* . . . Map Collectors' Circle No. 58 (London, 1969).

15. Faithorne, *Graveing*, p. 48.

16. Coolie Verner, "The Fry and Jefferson Map," *Imago Mundi* 21 (1967): 80.

17. Faithorne, *Graveing*, p. 43.

18. Ibid., p. 44.

19. Karel Kuchar, *Early Maps of Bohemia* (Prague, 1961), p. 27.

20. E. G. R. Taylor "Notes on John Adams and Contemporary Map Makers," *Geographical Journal* 97 (1941): 182–84.

21. Ibid., p. 183.

22. R. A. Skelton, *County Atlases of the British Isles* (London, 1970), p. 106.

23. Lewis Morris, Add. M.S. 604d, p. 58. National Library of Wales. Cited in A. H. W. Robinson, *Marine Cartography in Britain* (Leicester, 1962), pp. 81–82.

24. Verner, "Mr. Jefferson . . . ," p. 100.

25. Coolie Verner, "Surveying and Mapping the New Federal City," *Imago Mundi* 23 (1969): 66–67.

26. Arrowsmith, *Memoir*, p. 31.

27. Coolie Verner, *Maps By John Arrowsmith in the Publications of the Royal Geographical Society*. Map Collectors' Circle No. 76 (London, 1971).

28. Ibid., p. 11.

29. *Account* . . . , p. 166.

30. Faithorne, *Graveing*, p. 70.

31. Verner, "The Fry and Jefferson Map," p. 88. This ghost print is visible on the facsimile of the 1794 edition published in *The Fry and Jefferson Map of Virginia and Maryland: Facsimiles of the 1754 and 1794 Printings with An Index* (Charlottesville, 1966).

32. The chart of the Belt and Sound published in Mount and Page's edition of *The English Pilot. The Northerne Navigation* . . . in 1708 was printed from poorly cleaned plate that had been used originally for Seller's "A Chart of the Sea Coasts of Russia . . . ," ca. 1680. The ghost print of the original map is clearly visible even though the new map was cut upside down in relation to the original. This plate was used from 1708 to 1785.

33. Verner, "Mr. Jefferson . . . ," p. 102.

34. Hansard, *Typographia*, p. 801.

35. Faithorne, *Graveing*, pp. 61–62.

36. Ibid., p. 61.

37. Ibid., p. 66.

38. Hansard, *Typographia*, p. 802n.

39. Ibid., p. 803.

40. Faithorne, *Graveing*, p. 58.

41. Morris, add. M.S. 604d.

42. Verner, "Surveying and Mapping . . . ," p. 69.

43. Verner, "Mr. Jefferson . . . ," p. 106.

44. John F. M. Smith, *The Art of Painting in Oyl.* (London, 1769).

45. Faithorne, *Graveing*, p. 69.

46. Woodward, "Copper Engraving."

47. *Journal of the Royal Geographical Society* 15 (1845): cv.

48. See Tony Campbell, "The Drapers' Company and its school of seventeenth century chartmakers," in Helen Wallis and Sarah Tyacke, eds., *My Head is a Map* (London: 1973), pp. 81–106.

49. Arrowsmith, *Memoir*, p. 31.

50. Coolie Verner, *The English Pilot. The Fourth Book* . . . *A Carto-Bibliographical Study* (Charlottesville, 1960).

51. Henry G. Fordham, *John Cary* . . . (Cambridge, 1925), p. xxviii.

52. W. R. Chaplin, "A Seventeenth Century Chart Publisher," *American Neptune* 8 (1948): 302–4.

53. R. A. Skelton. *Decorative Printed Maps of the 15th to 18th Centuries* (London, 1952), p. 3.

54. Vittorio da Zonca, *Novo Teatro di Machine et Edificii* (Padua, 1607).

55. *Account*, p. 202.

56. R. A. Skelton. *County Atlases of the British Isles* (London, 1970), p. 36.

57. Coolie Verner, "Mark Tiddeman's Chart of New York Harbour," *American Neptune* 19 (January 1959): 44–50.

58. Verner, *The English Pilot, The Fourth Book,* pp. 33–35.

59. *Journal of the Royal Geographical Society* 12 (1842): lxxxvii.

60. Coolie Verner, *Maps of the Yorktown Campaign 1780–1781,* Map Collectors' Series No. 18 (London, 1965), p. 30, map 6.

61. This information is by courtesy of Tony Campbell.

62. Arthur H. Robinson, "Mapmaking and Map Printing: the Evolution of a working Relationship" (chap. 1 of this book).

63. This is amply illustrated in the confusion surrounding the application of terms and concepts from descriptive bibliography to cartobibliography. In so doing, the use of the term *edition* does not indicate whether a new plate is described or another printing from an old plate that has been altered. See Coolie Verner, "The Analysis of Variants in Maps Printed from Copperplates," *The American Cartographer* 1 (1974): 77–87.

64. Examples of research in this category are cited in the notes above.

65. See H. G. Fordham, *Studies in Carto-Bibliography* (Cambridge, 1914), and Edward Lynam, *The Mapmaker's Art* (London, 1953).

66. F. de Dainville, *Le Langage des Géographes* (Paris, 1964).

67. See, for example, Fordham, *John Cary;* A. S. Oshy, *Mercator,* (New York, 1969); J. Kenning, *Willem Jansz Blaeu,* revised and edited by M. Donkersloot-de Vrij, (Amsterdam, 1973).

4. Lithography and maps, 1796–1850

1. Alois Senefelder, *A Complete course of lithography* (London: Ackerman, 1819; reprinted, 1968, Da Capo Press, New York), p. 27.

2. Ibid., pp. 93–94.

3. Ibid., pp. 31–32.

4. Michael Twyman, *Lithography 1800–1850, the techniques of drawing on stone in England and France and their application in works of topography* (London: Oxford University Press, 1970), p. 64.

5. Senefelder, *complete course,* pp. 179–80.

6. Michael Twyman, "The lithographic hand press 1796–1850," *Journal of the Printing Historical Society* 3 (1967): 5. This article, one of the most authoritative on lithographic presses, includes illustrations of a number of early presses.

7. Ibid.

8. Ibid.

9. Luitpold Düssler, *Die Incunabeln der deutschen Lithographie* (1796–1821) (Heidelberg: Weissbach, 1955).

10. Senefelder, *complete course,* p. 63.

11. Ibid., pp. 276–77.

12. Twyman, *Lithography,* p. 64.

13. Ian Mumford, "Lithography, photography and photozincography in English map production before 1870," *The Cartographic Journal* 9 (1972): 31.

14. Ibid.

15. Ibid., p. 3.

16. Ibid.

17. Ibid., pp. 3–4.

18. "Curious specimen of polyautography, or lithography," *Gentleman's Magazine* 85 (1815): 297. This application was also reported by David Brewster, under "Lithography," in the *Edinburgh Encyclopaedia* 13 (1830): 44–47.

19. "Curious specimen . . . ," p. 297.

20. Elizabeth R. and Joseph Pennell, *Lithography and Lithographers* (New York: Macmillan, 1915), p. 37.

21. Ibid., p. 46.

22. Comte de Lasteyrie, "Rapport au nom d'une commission spéciale, sur les gravures lithographiques adressées à la Société par M. Engelmann," *Société d'encouragement* (Paris), *Bulletin* 14 (1815): 293.

23. Ibid., p. 294.

24. Walter Gräff, *Die Einführung der Lithographie in Frankreich* (Heidelberg: Rössler, 1906), p. 84.

25. Molly Fordham, "A note on maps and printing," *The Book-Collector's Quarterly* (London) 4 (Oct.-Dec. 1934): 77.

26. Twyman, "Lithographic hand press," p. 5.

27. Joseph Goodman, "The evolution of lithographic printing machinery," *Penrose Annual* (1904–05), p. 42.

28. Twyman, "Lithographic hand press," p. 22.

29. Ibid., pp. 49–50.

30. Kurt Krause, "150 Jahre reichsdeutsche Kartographie. Ein Beitrag zur Entwicklungsge-

schichte einiger privater kartographischer Anstalten," *Archiv für Buchgewerbe und Gebrauchsgraphik* 72 (1935):297–305.

31. Wilhelm Engelmann, *Bibliotheca Geographica, Verzeichniss der Seit der Mitte des vorigen Jahrhunderts bis zu Ende des Jahres 1856 in Deutschland erschienenen Werke über Geographie und Reisen mit Einschluss der Landkarten Pläne und Ansichten* (Leipzig, 1858).

32. Pennell, *Lithography*, p. 53.

33. Jomard [Edme], *Rapporteur.* "Examen de la question de savoir: 1. Si la lithographie peut-être appliquée avec avantage à la publication des cartes géographiques, tant sous le rapport du mérite de l'exécution que sous le rapport de l'économie? 2. Jusqu'à quel point elle peut remplacer, pour cet objet, la gravure sur cuivre?" Société de géographie (Paris), *Bulletin* 5(1826): 693–94.

34. Ibid., p. 697.

35. Ibid., p. 707.

36. Ibid., p. 708.

37. Ibid.

38. *Le Lithographe, Journal des Artistes et des Imprimeurs,* dirige par M. Jules Desportes, 4(1845):314.

39. Ibid.

40. For an excellent summary of Santarém's several facsimile atlases see Armando Cortesão, *History of Portuguese Cartography* (Coimbra, 1969) 1:7–22.

41. Ibid., pp. 17–18.

42. Henri M. A. Berthaut, *La carte de France 1750–1898* (Paris: 1897–99) 2:104.

43. Twyman, *Lithography*, p. 37.

44. Ida Darlington and James Howgego, *Printed maps of London circa 1553–1850* (London, 1964), p. 36.

45. Ibid., p. 37.

46. Archibald Day, *The Admiralty Hydrographic Service 1795–1919* (London, 1967), p. 36.

47. *One hundred years of map making, the story of W. & A. K. Johnston* (Edinburgh: W. & A. K. Johnston, 1926), pp. 9–10.

48. George Philip, *The story of the last hundred years, a geographical record* (London: George Philip & Son, 1934), p. 49.

49. Benjamin P. Wilme, *A hand book for plain and ornamental mapping,* (London: Weale, 1846), pl. a.

50. Ibid., p. 57.

51. William Smith Williams, "On lithography," Society of Arts (London), *Transactions* (May 1849), part II, p. 232.

52. Ibid., p. 246.

53. Mumford, "Lithography,...," p. 32.

54. Society for the Promotion of Christian Knowledge, *The History of Printing* (London, 1862), p. 214. For a more complete discussion on this subject see Walter W. Ristow, "The anastatic process in map reproduction," *The Cartographic Journal* 9 (1972):37–42.

55. Personal communication from Professor C. Koeman, 5 May 1972.

56. E. Gilbert de Cauwer, "Philippe Vandermaelen, Belgian geographer, 1795–1869," *Imago Mundi* 24 (1970):11–20.

57. Ibid., p. 13.

58. Bundesamt für Eich- und Vermessungswesen, *Die amtliche Kartographie Österreichs* (Vienna, 1970), p. 21.

59. Ernst Nischer von Falkenhof, *Österreichische Kartographen, Ihr Leben, Lehren und Wirken* (Vienna, 1925), p. 162.

60. Bundesamt, *Kartographie Österreichs*, p. 28.

61. Cauwer, *Vandermaelen*, p. 14.

62. Rollo G. Silver, *The American printer 1787–1825* (Charlottesville: Bibliographical Society of the University of Virginia, 1967), p. 61.

63. Peter C. Marzio, "American lithographic technology before the Civil War," in: John D. Morse, ed., *Prints in and of America to 1850* (Charlottesville: The University Press of Virginia, 1970), p. 215.

64. Ibid., p. 216.

65. Ibid., p. 221.

66. "Notice of the lithographic art," *American Journal of Science and Arts* 4 (1822):169.

67. Charles B. Wood, III, "Prints and scientific illustration in America," in: Morse, *Prints,* p. 178.

68. Cadwallader D. Colden, *Memoir, prepared at the request of a committee of the Common Council of the City of New York, and presented to the Mayor of the City, at the celebration of the completion of the New York canals* (New York, 1825), p. 349.

69. Ibid., p. 354–55.

70. Hans F. Bauman, *Artists' lithographs: a world history from Senefelder to the present day* (London, 1970), p. 44. George C. Groce and David H. Wallace, *Dictionary of artists in America 1564–1860* (New Haven: New-York Historical Society, 1957), p. 607. Silver, *American Printer*, p. 169.

71. Silver, *American Printer*, p. 169.

72. Christopher Columbus Baldwin, *Diary* (Worcester, Mass., 1901), p. 331–32.

73. John Thomas Carey, "The American lithograph from its inception to 1865 with biographical considerations of twenty lithographers and a checklist of their works," dissertation, Ohio State University, 1954, p. 79.

74. "Lithography," *Boston Monthly Magazine* (Dec. 1825): 383–84.

75. Harry T. Peters, *Currier and Ives: printmakers to the American people* (New York: Doubleday, 1929–31), 1:5.

76. As quoted by J. D. Whitney, "Geographical and geological surveys," *North American Review* 121 (1875):77.

77. Simeon Borden, "Account of a trigonometrical survey of Massachusetts, with a comparison of its results with those obtained from astronomical observations by Robert Treat Paine," American Philosophical Society, *Transactions* 9 (1846):34.

78. Charles Taylor, "Some notes on early American lithography," American Antiquarian Society, *Proceedings,* 32 (1922):75.

79. Peters, *Currier and Ives, 1:*5.

80. Jane Cooper Bland, *Currier & Ives, a manual for collectors* (Garden City, N. Y., 1939).

81. Quoted in W. S. Hughes, *Founding and development of the U. S. Hydrographic Office* (Washington, 1887), p. 8.

82. Ibid., p. 9.

83. Senefelder, *complete course,* p. 256.

84. Colden, *Memoir,* p. 350.

85. For more details on this see Ristow, "The anastatic process . . .," and his "The map publishing career of Robert Pearsall Smith," *Quarterly Journal of the Library of Congress* 26 (1969):170–96.

86. See Ristow, "Robert Pearsall Smith," pp. 184–93.

5. Miscellaneous map printing processes in the nineteenth century

1. Used by Edward Palmer of London in the 1840s. Similar to the German "Galvanographie."

2. Louis Schoenberg, *Metallic engravings in relief for letterpress printing . . .,* Part 3 (London, 1842).

3. Siderography was developed from a number of his own inventions by the engineer Jacob Perkins of Philadelphia and adopted by some American banks after 1810 and English banks after 1819. See Greville and Dorothy Bathe, *Jacob Perkins. His inventions, his times and his contemporaries* (Philadelphia, 1943).

4. Heinz Bosse, *Kartentechnik* (Gotha, 1951), 2:33. Ergänzungsheft Nr. 245 *Petermanns Geographische Mitteilungen.*

5. Denis Diderot, *Encyclopédie méthodique.* (Paris: Pancouke, 1787–1832) *Arts et métiers mécaniques* 3:248 gravure and two plates.

6. Omnigraphy is described in William J. Stannard, *The Art Exemplar* (London, 1859), p. 186; and two examples of omnigraphic maps are in the Smithsonian's Warshaw Collection.

7. *The American Bookmaker* 6 (1888):130.

8. Two monographs on the composed map are Douglas C. McMurtrie, *Printing geographic maps with movable types* (New York, 1924), and Edouard Hoffman-Feer, *Die Typographie im Dienste der Landkarte* (Basel, 1969).

9. J. G. I. Breitkopf, *Über den Druck der geographischen Charten* (Leipzig, 1777).

10. Hoffman-Feer, *Typographie,* p. 6.

11. Hoffman-Feer, *Typographie,* p. 47.

12. Brevet d'Invention no. 2216, 25 September 1823.

13. Reference from McMurtrie, *Printing,* pp. 14–16, where the specimen page is reproduced.

14. A description and bibliography of the process is in Elizabeth Harris, "Experimental graphic processes in England 1800–1859," *Journal of the Printing Historical Society* 4 (London, 1968):74.

15. Thomas B. van Horne, *History of the Army of the Cumberland* (Cincinnati, 1875).

16. There were also a few etched black-line relief processes, such as . . . William Blake's, Charles Pye's (*London Journal* 1 (1820): 55, 78), or W. H. Lizar's (*Edinburgh Philosophical Journal* 2 (1820):19), but I am not aware that these were used for maps.

17. For example, "Engraving in stereotype" reported in *Mechanics' Magazine* 16 (1832): 334, and Oxley's process reported in *Mechanics' Magazine* 29 (1838):163.

18. *Gentleman's Magazine,* London, new series 2 (1834):250.

19. Edward Palmer: British Patent No. 8987, 1842, and No. 9227, 1843. Edward Palmer: *Glyphography, or engraved drawing, for printing at the type press after the manner of woodcuts* (London, 1st and 2d eds., 1843, 3d ed., 1844).

20. W. A. Blackie, *The Imperial Gazetteer* (London, 1855).

21. Alfred Smee, *Elements of Electrometallurgy* (London, 1851, 3d ed.), p. 324.

22. *Reports of the Superintendent of the Coast Survey* (Washington, D. C.).

23. Ibid., 1856, p. 155.

24. Ibid., 1857, p. 211.

25. *Photolithography. Report of the Board Appointed by the Hon. the President of the Board of Land and Works . . .* (Melbourne, 1860–61).

26. Ibid., p. 11, question 253.

27. *Journal of the Society of Arts* 9 (1860):39.

28. *London Journal* 1 (1820):388.

6. The application of photography to map printing and the transition to offset lithography

1. For literature on the history of photography see: Peter Pollack, *The Picture History of*

Photography, rev. ed. (New York, 1970); H. and A. Gernsheim, *The History of Photography* (New York, 1969); J. M. Eder, *Geschichte der Photographie* (Halle, 1932), trans. by E. Epstein, *History of Photography* (New York, 1945).

For general information on graphic techniques, the author consulted N. G. van Huffel, *Encyclopedisch Handboek der Graphische werkwijzen* (Utrecht, 1926).

2. I am most grateful for this information, communicated to me by Monsieur Pognon of the Bibliothèque Nationale, Paris.

3. W. F. Versteeg, "Vermenigvuldiging van kaarten" (Map-reproduction), *Tijdschrift van het Aardrijkskundig Genootschap* 3 (1879):4–21.

4. Ian Mumford, "Lithography, Photography and Photozincography in English Map Production before 1870," *Cartographic Journal* 9 (1972): 30–36. An interesting account of the relationship between James and Osborne is to be found in the preface to A. de C. Scott, *On Photo-Zincography* (2d ed.; London: Longman, 1863).

5. O. Volkmer, *Die Technik der Reproduktion von Militärkarten und Plänen nebst Ihrer Vervielfältigung* (Vienna, 1885). Volkmer mentions the following publications: Ordnance Survey, *Methods and Processes Adopted for Production of the Maps of the Ordnance Survey* (1875). J. Waterhouse, *The Application of Photography to the Reproduction of Maps and Plans by the Photomechanical and Other Processes* (London, 1878). J. Rodrigues, *La Section photographique et artistique de la direction générale des travaux géographiques du Portugal* (Lisbon, 1877).

6. Contemporary literature: *Le Lithographe, Journal des artistes et des imprimeurs, publiant tous les procédés connus de la lithographie* ... (Paris, 1838). G. Fortier, *La Photolithographie, son origine des procédés, ses applications* (Paris, 1876). History of lithography: R. Graul and Fr. Dörnhöffer, *Die Lithographie von ihrer Erfindung bis zur Gegenwart* (Vienna, 1903). A. Lemercier, "La Lithographie française de 1796–1896." (Paris, Ch. Lorillieux & Cie, 1900). M. Twyman, *Lithography, 1800–1850* (London, 1970).

7. W. W. Ristow, "The Anastatic Process in Map Reproduction," *Cartographic Journal* 9 (1972): 37–42.

8. Mumford, "Lithography," p. 32.

9. M. Huguenin, *Historique de la cartographie de la nouvelle Carte de France*. Publication technique de l'Institut Géographique National (Paris, 1948), p. 5.

10. *Een methode van kleuren- en plaatdruk door het etsen van tinten op steen en de typo-autographie*. Uitgegeven door het Departement van Oorlog (The Hague, 1878).

11. F. de Bas, *Die Residentie-kaarten van Java en Madoera*. Uitgegeven van wege het Aardrijkskundig Genootschap (Amsterdam, 1876).

12. *Aanteekeningen omtrent de geschiedenis en de inrichting der Waterstaatskaart van Nederland op de schaal van 1:50.000* (The Hague, 1892).

13. A. Bortzell, *Beskrifning öfver Besier-Ecksteins Kromolitographi och Litotypografi använde vid trykningen of geologisk ofversigtskarta öfver Skäne* (Stockholm, 1872).

14. Ch. Eckstein, *New Method for Reproducing Maps and Drawings*. International Exhibition at Philadelphia (The Hague, 1876).

15. C. Vogel, "Die Kartographie auf der Pariser Weltausstellung 1878," *Petermanns Geographische Mitteilungen* 24 (1878): 445–60.

16. *Annual Report of the Director of the U.S. Geological Survey, 1888–1889* (Washington, 1889).

17. See the following works by O. Volkmer: *Die Technik der Reproduktion von Militär-Karten und Plänen des K. K. Militär-Geographischen Instituts zu Wien.* (Separat-beilage zu den "Mittheilungen über Gegenstände des Artillerie- und Genie-Wesens") (Vienna, 1880). *Die Technik der Reproduktion von Militärkarten und Plänen nebst Ihrer Vervielfältigung* (Vienna, 1885). "Die Verwerkung der Elektrolyse in den graphischen Künsten," *Mittheilungen des Militärgeographischen Instituts Wien* 4 (1884): 79.

18. Versteeg, "Vermenigvuldiging van kaarten," p. 19, 20.

19. Edme Jomard, *Examen de la question de savoir: 1. si la lithographie peut-être appliquée avec avantage à la publication des cartes géographiques tant sous le rapport du mérite de l'exécution que sous le rapport de l'économie. 2. jusqu'à quel point elle peut remplacer cet objet, la gravure sur cuivre* Société de Géographie, *Bulletin* 5 (1826): 693–94.

20. A. Petermann, "Die Sonne im Dienste der Geographie und Kartographie. Die Sonnen-Kupferstich (Heliogravüre) und die neue Generalstabskarte der Österreichische-Ungarischen Monarchie in 715 Blättern," *Petermanns Geographische Mittheilungen* 24 (1878): 205–13.

21. O. H. Krause, *Neue Wege der Kartenherstellung im Reichsamt für Landesaufnahme*. Sonderheft 9 zu den "Mitteilungen des Reichsamts für Landesaufnahme" (1935).

22. H. Mühle, "Die Drucktechnik, ihre Entwicklung in der Kartographie der letzten 25 Jahre," *15 Jahre Ortsverein Stuttgart der Deutschen Gesellschaft für Kartographie* (Stuttgart, 1968).

23. T. A. Wilson, *The Practice of Collotype* (London, 1935).

24. F. Schrader, *Atlas de géographie moderne* (Paris, 1895). See Carte no. 9: "La France et les régions environnantes."

25. G. Fritz, *Die Photolithographie* (Halle, 1894). Fr. Hesse, *Die Chromolithographie, mit besonderer Berücksichtigung der modernen, auf photographischer Grundlage beruhenden Verfahren und der Technik des Aluminiumdrucks* (2d ed.; Halle, 1906).

26. F. Hesse, "Die Reproduktion von Karten und Plänen," *Klimsch Jahrbuch* 6 (1905): 89–124.

27. Mühle, "Die Drucktechnik," p. 74.

28. A. Krämer, "Zeichenplatten aus Zelluloid," *Allgemeine Vermessungsnachrichten* 13 (1929): 28.

29. W. Kleffner, *Die Reichskartenwerke* (Berlin, 1939).

30. D. Chervet, "Verfahren zur Herstellung mehrfarbigen Karten." *Fortschrittsberichte auf dem Gebiet des Kartendrucks und Karten reproduktion* (Erster Intern. Kurs für Kartendruck und Reproduktion, 1956, Munich; Hamburg, 1956).

31. D. Woodward, "A Note on the History of Scribing," *Cartographic Journal* 3 (1966): 58. Reference is made to the manual by John Jacob Holtzapffel, *Turning and Mechanical Manipulation*. vol. 5: *The Principles and Practice of Ornamental or Complex Turning* (London, 1884).

32. B. A. Snissarenko, "Anwendung der Gravierung auf Glas in der Kartographie," *Geodezist* 6 (1939).

33. Th. J. Ouborg, "Het reproductiebedrijf," *75 Jaren Topografie in Nederlandsch-Indië* (Batavia, 1939): 63.

34. See the bibliography of cartography in H. P. Kosack and K. H. Meine, *Die Kartographie 1943–1954* (Lahr/Schwarzwald: Astra Verlag, 1955), (*Kartographische Schriftenreihe*, Bd. 4) and *Bibliotheca Cartographica* (Selbstverlag der Bundesanstalt für Landeskunde, Bad Godesberg, 1957–); for literature on cartography over the period 1866–1943 see *Geographisches Jahrbuch* (Gotha: Justus Perthes).

35. H. Knorr, "Spezielle fotomechanische Verfahren in der modernen Kartenherstellung." (Erster Intern. Kurs für Kartendruck und Reproduktion, 1956, Munich; (Hamburg, 1956): 95–106. F. A. Clemens, "Electrophotography. Its Impact on Cartography." *Nachrichten aus dem Karten- und Vermessungswesen,* Reihe V, Heft No. 5 (Frankfurt-am-Main, 1963): 109–13.

36. "Esselte Conference on Applied Cartography. Stockholm, July 1956," *Nachrichten aus dem Karten- und Vermessungswesen,* Reihe II, Heft 2. (Frankfurt-am-Main, 1958): 1–85.

37. "Second International Cartographic Conference, Chicago, June 1958," *Nachrichten aus dem Karten- und Vermessungswesen,* Reihe II, Heft 3–7 (Frankfurt-am-Main, 1959). Also published in Reihe I, Heft 7–11. (Sammelband, in German)

Selected bibliography

This bibliography includes works of two categories: (a) books and articles containing general material relating to the history of map printing, in which more specific references may be found, and (b) some of the more unusual articles that may not be found listed in the common sources. For further references, the reader is referred to the Library of Congress Geography and Map Division *Bibliography of Cartography* 4 vols. (Boston: G. K. Hall, 1973), the continuing *Bibliotheca Cartographica* (Bonn-Bad Godesberg, 1957–) and the annual bibliography in *Imago Mundi* (London, 1935–). A useful guide to the various editions of early printing manuals is found in Levis (1912).

Abbadie, Antoine Thompson d'. "Sur la gravure des cartes." *Société de Géographie Commerciale de Bordeaux. Bulletin* [2. serie,] 5 (1882): 141–43.

Albach, Julius. *Der Einfluss der modernen Reproductions-mittel auf die Kartographie.* Vienna, 1876.

Amann, Joseph. *Die bayerische Landesvermessung in Ihrer geschichtlichen Entwicklung. Erster Teil: Die Aufstellung des Landes vermessungwerkes, 1808–1871.* Munich: K. B. Katasterbureau, 1908.

Anesi, Jose. "Aspectos Graficos de la Cartografia." *Revista Geographica Americana* 30 (1948): 169–84.

Bagrow, Leo. *History of Cartography,* rev. and enlarged by R. A. Skelton. London: C. A. Watts, 1964.

Baltimore Museum of Art. *The World Encompassed.* An exhibition of the *History of Maps Held at the Baltimore Museum of Art. October 7–November 23, 1952.* Baltimore: Trustees of the Walters Art Gallery, 1952.

Bay, Helmuth. *The History and Technique of Map Making.* New York: New York Public Library, 1943 [no. 8, Richard Rogers Bowker Memorial Lectures].

Bigmore, E. C., and Wyman, C. W. H. *A Bibliography of Printing with Notes and Illustrations.* London: Bernard Quaritch, 1880; reprinted London: Holland Press, 1969.

Blaikie, W. B. "How Maps are Made." *Scottish Geographical Magazine* 7 (1891): 419–34.

Bosse, Abraham. *Traicté des Manières de Graver.* Paris, 1645.

Bosse, Heinz. "Kartentechnik—II Vervielfältigungsverfahren." *Erganzungsheft Nr. 245 zu Petermanns Geographische Mitteilungen.* Gotha: Justus Perthes, 1951.

Breitkopf, Johann Gottlob Immanuel. *Ueber den Druck der Geographischen Charten. Nebst Beygefugter Probe Einer Durch die Buchdruckerkunst Gesetzten und Gedruckten Landcharte.* Leipzig: Breitkopfischen Buchdruckerey, 1777.

British Museum. *A Guide to the Processes and Schools of Engraving. . . .* London, 1914.

Brown, Lloyd A. *The Story of Maps.* Boston; Little, Brown and Co., 1949.

Camus, Armand Gaston. *Mémoire sur l'Impression des Cartes Géographiques, etc.* Paris, 1798. "Les cartes photolithographiées." *Journal général de l'imprimerie et de la librairie. Cronique* [2 serie,] 17 (1873): 139.

Cayley, Henry. "On the Colouring of Maps." *Proceedings of the Royal Geographical Society* 1 (1879): 259–61.

Chatto, William Andrew. *A Treatise on Wood Engraving, Historical and Practical.* 2d ed. London: Charles Knight. 1861.

Crone, G. R. "The Evolution of Map Printing." *Printing—A Supplement Published by The Times on the Occasion of the 10th International Printing Machinery and Allied Trades Exhibition.* London, 1955.

Dainville, François de. *Le Langage des Géographes.* Paris: A. and J. Picard, 1964.

Delen, Adrien Jean Joseph. *Histoire de la gravure dans les anciens Pays-Bas et dans les provinces belges, des origines jusqu'à la fin du xviiie siècle.* Paris: G. Van Oest, 1924.

Dembour, A. *Description d'un Nouveau Procédé de Gravure en Relief sur Cuivre, Dite Ectypographie, Inventé par A. Dembour.* Metz: S. Lamort, 1835.

Diderot, Denis. *Encyclopédie, ou Dictionnaire Raisonné des Sciences.* Paris: Briasson, etc., 1751–65.

Didot, M. Ambroise Firmin. "Cartes géographiques." *L'Imprimerie la librarie et la papeterie; l'exposition universelle de 1851.* Paris: Imprimerie Impériale ,1854, pp. 38–40.

Düssler, Luitpold. *Die Incunabeln der deutschen Lithographie (1796–1821).* Heidelberg: Weissbach, 1955.

Eckert, Max. *Die Kartenwissenschaft.* Berlin & Leipzig: Walter de Gruyter & Co., 1921.

Eckstein, Charles. *New Method for Reproducing Maps and Drawings.* The Hague: Topographical Department, War Office and Giunta d'Albani Borthers [International Exhibition at Philadelphia, 1876].

Faithorne, William. *The Art of Graveing and Etching* . . . 2d ed. London: A. Roper, 1702.

Fordham, Molly. "A note on maps and printing." *The Book-Collector's Quarterly* (London) 4 (Oct.—Dec. 1934): 77.

Galton, Francis. "Geography." [An address on British Ordnance Survey maps, color printing, and cartography.] British association for the advancement of science. Report of the 42d meeting, 1872. London, 1873. Pp. 198–203.

Grenacher, Franz. "The Woodcut map." *Imago Mundi* 24 (1970): 31–41.

Hind, Arthur M. *An Introduction to a History of Woodcut, with a Detailed Survey of Work Done in the Fifteenth Century.* Boston: Houghton Mifflin Co., 1935; reprinted London: Constable & Co., 1963, and New York: Dover Publications, 1963.

———. *A History of Engraving and Etching from the 15th Century to the Year 1914.* Boston: Houghton Mifflin Co., 1923; reprinted New York: Dover Publications, 1963.

Hodson, James Shirley. *An Historical and Practical Guide to Art Illustration, in Connection with Books, Periodicals and General Decoration.* London: Sampson Low, Marston, Searle & Rivington, 1884.

Hoffman-Feer, Eduard. *Die Typographie im Dienste der Landkarte.* Basel: Helbing & Lichtenhahn, 1969.

Horn, Werner. "Sebastian Münster's Map of Prussia and the Variants of It." *Imago Mundi* 7 (1951): 66–73.

Huck, Thomas William. "The Earliest Printed Maps." *The Antiquary* 46 (1910): 253–57.

Huguenin, M. *Historique de la cartographie de la nouvelle Carte de France.* Publication technique de l'Institut Géographique National. Paris, 1948.

Hupp, Otto. *Philipp Apian's Bayerische Landtafeln und Peter Weiner's-Chorographia Bavariae, Eine Bibliographische Untersuchung.* Frankfurt am Main: Heinrich Keller, 1910.

Ivins, William M., Jr. *Prints and Visual Communication.* Cambridge, Mass.: Harvard University Press, 1953.

Jomard, M. "Examen de la question de savoir: 1. Si la lithographie peut-être appliqué avec avantage à la publication des cartes géographiques, tant sous le rapport du mérite de l'exécution que sous le rapport de l'économie? 2. Jusqu'a quel point elle peut remplacer, pour cet objet, la gravure sur cuivre?" *Bulletin de la Société de Géographie de Paris* [1. série] 5 (1826): 693–708.

Koppe, C. "Wesen und Bedeutung der 'graphischen Kunsts' für den Illustrations—und Kartendruck." *Himmel und Erde* 10 (1898): 193–211, 268–82.

Legros, Lucien Alphonse, and Grant, John Cameron. *Typographical Printing-Surfaces, The Technology and Mechanism of Their Production.* London: Longmans, Green. 1916.

Levis, Howard C. *A Descriptive Bibliography of the most important books in the English language relating to the Art and History of Engraving and the Collecting of Prints.* London: Ellis, 1912.

Lister, Raymond. *How to Identify Old Maps and Globes.* London: G. Bell and Sons, 1965.

Lynam, Edward. *The Mapmaker's Art.* London: Batchworth Press, 1953.

McMurtrie, Douglas C. *Printing Geographic Maps with Movable Types.* New York: privately printed, 1925.

————. *Stereotyping in Bavaria in the Sixteenth Century—A Note on the History of Map Printing Processes and of Printer's Platemaking.* New York: privately printed, 1935.

Mathiot, George. "On the Electrotyping Operations of the U. S. Coast Survey." *American Journal of Science and Arts* [2d series] 15 (1853): 305–9.

Mayer, Max. "Maps and Their Making." *The Publishers' Weekly* 117 (1930): 2,841–47; 118 (1930): 70–78, 973–76, 1,663–66.

Motte, Philip, H. de la. *On the Various Applications of Anastatic Printing and Papyrography with Illustrative Examples.* London: David Bogue, 1849.

Mumford, Ian. "Lithography, photography and photozincography in English map production before 1870." *The Cartographic Journal* 9 (1972): 30–36.

Nebenzahl, Kenneth. "A Stone Thrown at the Map Maker." *The Papers of the Bibliographic Society of America* 55 (1961): 283–88.

"A New Way of Printing Maps." *The American Lithographer and Printer* (1886): 126.

New Zealand Government. Lithographic Department. "Photolithography as Practised at the New Zealand Government Lithographic Department." Wellington, N.Z., 1875.

[Niblach, A. P.] "Chart plates; engraving, correcting and repair." *Hydrographic Review* 1, 2. (1924): 27–37.

Noirot, A. "La gravure sur pierre appliquée aux cartes." *Spectateur militaire* [2 serie] 49 (1865): 156–60.

Oldfield, Margaret. "Five Centuries of Map Printing." *The British Printer* 63 (1955): 69–74.

Ordnance Survey. *Methods and Processes Adopted for Production of the Ordnance Survey.* London, 1875.

————. *Account of the Methods and Processes Adopted for the Production of the Maps of the Ordnance Survey.* 2d ed. London: H. M. Stationery Office, 1902.

Osley, A. S. *Mercator: A Monograph on the Lettering of Maps. . . .* New York: Watson-Guptill, 1969.

"Other Work . . . in the Printing Department of the Ordnance Survey Office. . ." *The Printing Times and Lithographer* 9 (1883): 111.

Papillon, J. R. M. *Traité Historique et Pratique de la Gravure en Bois.* 3 vols. Paris: Pierre Guillaume Simon, 1766.

Pearson, Karen. "Lithography and Maps in Nineteenth-Century Geographical Periodicals," unpublished paper in progress, Department of Geography, University of Wisconsin, Madison.

Prideaux, S. T. *Aquatint Engraving. . . .* London: Duckworth and Co., 1909.

"The Printer's Devil." *Quarterly Review* 65 (December 1839): 28. (Map printing method of Charles Knight.)

"Printing Abroad. Austria. War Maps." *The Printing Times and Lithographer* 2 (1876): 194.

Remius, C. P. "Some results of experiments in chart printing from copper plates." *Hydrographic Review* 6 (1929): 29–38.

Ridgway, John Livesy. *The Preparation of Illustrations for Reports of the United States Geological Survey, with Brief Descriptions of Processes of Reproduction.* Washington, D.C.: Government Printing Office, 1920.

Ristow, Walter W. "The anastatic process in map reproduction." *The Cartographic Journal* 9 (1972): 37–42.

Ritschel von Hartenbach, J. *Neues System Geographische Charten zugleich mit Ihrem Colorit auf der Buchdruckerpresse herzustellen.* Leipzig, 1840.

Rodrigues, J. *La Section photographique et artistique de la direction générale des travaux géographiques du Portugal.* Lisbon, 1877.

Salmon, William. *Polygraphice; or, the arts of drawing, engraving.* London 1675 (5th ed. 1685—chapter "of colors simple for washing of maps—of colors compounded for washing maps," pp. 204–11).

Schikofsky, Karl. *Reproduktions-methoden zur herstellung von Karten.* Vienna: L. W. Siedel, 1890.

Scott, A. de C. *On Photo-Zincography and other Photographic Processes,* 2d ed. London: Longman, Green, Longman, Roberts & Green, 1863.

Senefelder, Alois. *A Complete Course of Lithography Containing Clear and Explicit Instructions in all the Different Branches and Manners of that Art: . . . a History of Lithography. . . .* London: R. Ackermann, 1819.

————. *The Invention of Lithography.* [Translated from the German by J. W. Muller.] New York: Fuch and Lang Manufacturing Company, 1911.

Skelton, R. A. "Cartography." In Singer, Charles, et al., eds. *A History of Technology,* vol. 4, pp. 596–628. New York and London: Oxford University Press, 1958.

————. "Colour in Mapmaking." *The Geographical Magazine* 32 (1960): 544–53.

————. "Decoration and Design in Maps before 1700." *Graphis* 7 (1951): 400–13.

————. *Decorative Printed Maps of the 15th to 18th Centuries.* London and New York: Staples Press, 1952.

————. "Early Maps and the Printer." *Printing Review* 16 (1951): 39–40.

————. "The Early Map Printer and His Problems." *Penrose Annual* 57 (1964): 171–84.

Thordeman, Bror. "Om kartreproduktion." *Globen* 4 (1925): 23–31.

Twyman, Michael. *Lithography 1800–1850, the techniques of drawing on stone in England and France and their application in works of topography.* London: Oxford University Press, 1970.

Volkmer, Ottomar. *Die Technik der Reproduction von Militär-Karten und Plänen nebst ihrer Vervielfältigung.* Leipzig, 1855.

————.*Die Technik der Reproduction von Militär-Karten und Plänen des K. K. Militär-Geographischen Instituts zu Wien.* [Separat-beilage zu den "Mittheilungen uber Gegenstande des Artillerie-und Genie-Wesens"] Vienna, 1880.

Wagner, Eduard. "Kartographische Reproduktionsmethoden." *Geographische Zeitschrift* 15 (1909): 204–19.

Wakeman, Geoffrey. "Lithography, photography and map printing," in *Aspects of Victorian Lithography.* Wymondham: Brewhouse Press, 1970, pp. 35–40.

————. *Victorian Book Illustration. The Technical Revolution.* Detroit: Gale Research Company, 1973.

Waterhouse, J. *The Application of Photography to the Reproduction of Maps and Plans by the Photomechanical and Other Processes.* London, 1878.

Wilme, Benjamin P. *A hand book for plain and ornamental mapping.* London: Weale, 1846.

Wood, H. Trueman. *Modern Methods of Illustrating Books.* 3d ed. London: Elliot Stock, 1890.

Zonca, Vittorio da. *Novo Teatro di Machine et Edificii.* Padova: 1607.

Index